Contamination Control in
Trace Element Analysis

CHEMICAL ANALYSIS

A SERIES OF MONOGRAPHS ON
ANALYTICAL CHEMISTRY AND ITS APPLICATIONS

Editor

P. J. ELVING

Editor Emeritus: **I. M. KOLTHOFF**

VOLUME 47

A WILEY-INTERSCIENCE PUBLICATION

JOHN WILEY & SONS
New York / London / Sydney / Toronto

Contamination Control in Trace Element Analysis

MORRIS ZIEF

*Principal Scientist for Purification
Research & Development, J. T. Baker Chemical Company,
Phillipsburg, New Jersey*

JAMES W. MITCHELL

*Head, Analytical Research Department,
Bell Telephone Laboratories,
Murray Hill, New Jersey*

A WILEY-INTERSCIENCE PUBLICATION

JOHN WILEY & SONS
New York / London / Sydney / Toronto

Library of Congress Cataloging in Publication Data:

Zief, Morris.
  Contamination control in trace element analysis.

  (Chemical analysis; v. 47)
  "A Wiley-Interscience publication."
  Includes bibliographical references and index.
  1. Trace elements—Analysis. 2. Contamination
(Technology) I. Mitchell, James W., joint author.
II. Title. III. Series.
QD139.T7Z5   545   76-16837
ISBN 0-471-61169-7

Printed in the United States of America

10 9 8 7 6 5 4 3 2 1

TO OUR WIVES, JEAN AND HAZEL

PREFACE

The supply-demand relationship of the authors' employers, a commercial supplier of high purity and specialty chemicals and an advanced materials research laboratory, led to the development of an interdisciplinary team for the production and characterization of ultrapure chemicals for communications technology. During several years of collaboration it was clear that success in the joint venture depended on the selection of purification methods, handling techniques, and reliable procedures for monitoring purification processes and analyzing final products. This experience convinced the authors that a book on practical contamination control techniques specifically for ultratrace analysis would be a valuable reference for a wide range of technologists and scientists. Thus the primary objective is the presentation of a state-of-the-art review of contamination control, specifically applicable to trace analysis, but also useful in all aspects of ultrapurity work.

In Chapter 1 we define purity and the parameters that limit the accuracy of ultratrace measurements. Chapter 2 discusses the problems most frequently encountered in quantitative, ultratrace analysis. Here the intent is to increase the analyst's level of awareness and recognition of pitfalls. In Chapter 3 information for the design and construction of an economic, clean analytical facility is provided. Chapter 4 covers the selection and cleaning of containers and apparatus. Chapter 5 reviews the laboratory methods most useful for purifying analytical reagents. Application of the methods rather than theory is emphasized. Throughout these chapters we give sources of specialized items helpful in ultratrace work.

Chapter 6 deals with methodology and techniques for contamination control during routine laboratory procedures. The last chapter contains both moderately detailed and cursory discussions of trace analytical techniques suitable for high accuracy work. Some methods have been given detailed treatment even though their application might be limited to a few elements, or expertise is available in relatively few existing analytical laboratories. Other techniques more widely used for economic reasons or for broad applicability rather than accuracy receive less attention. Applications of the methods in practical quantitative characterization of high purity chemicals are stressed. It was not our purpose to discuss the theory, principles, and instrumentation pertaining to all cases, since these topics are adequately described elsewhere.

Although we are reasonably sure that our primary goals have been attained during this writing, the degree to which we have accomplished the stated purposes and objectives can be determined only by the reader. We welcome suggestions, comments, corrections, and criticism.

We take pleasure in acknowledging the contribution on gases in Chapter 5 by Mr. C. A. McMenamy, J. T. Baker Chemical Company. We also are grateful for the assistance and encouragement provided by the Bell Telephone Laboratories and the J. T. Baker Chemical Company.

JAMES W. MITCHELL
MORRIS ZIEF

Phillipsburg, New Jersey
Murray Hill, New Jersey
April 1976

CONTENTS

Contamination Control in
Trace Element Analysis

CHAPTER 1

INTRODUCTION

I. TRACE ANALYSIS

A. ROLE OF CONTAMINATION

Contamination is one of the major problems hampering analyses at the trace (1–100 $\mu g/g$) and ultratrace (<1 $\mu g/g$) level, as illustrated by the literature on the analysis of seawater. Even though 8 trace and more than 40 ultratrace elements are distributed fairly uniformly in the ocean, extremely large variations in elemental abundance have been reported. As sampling and analytical methods have been improved, the concentrations of many elements in seawater have been found to be substantially lower, some by orders of magnitude. Another classic example of the effects of contamination is the recent controversy concerning polywater, a "new" species of water. At one time several theoreticians offered quantum mechanical diagrams for the bonding required to explain the anomalous properties. Scientists were misled, however, because they had disregarded small amounts of trace impurities. Hindsight has shown that polywater is really nothing more than polycontamination.

Scientists carefully considered the effects of contamination during the analysis of lunar samples. Soon after the first lunar samples were returned to the earth by the Apollo 11 astronauts on July 24, 1969, investigators throughout the world became part of a historic, coordinated analytical team. For months prior to the scheduled Apollo flight, chemists and geologists had been preparing contamination-free hydrochloric, hydrofluoric, and phosphoric acids for the dissolution of the geological specimens. The *accurate* determination of the age of the moon by isotope-abundance studies was highly dependent on lead-free reagents.[1] In spite of all possible care to preserve the integrity of the samples, definite contamination by fluorine, silver, and indium was recognized; possible contamination by nitrogen was also suggested.[2] The fluorine was derived from Teflon bags in which the astronauts sealed the samples on the moon. Microscopic examination by one research team revealed the presence of translucent Teflon fibers abraded from the plastic container.[3] Traces from the indium-silver O-ring that sealed the outer sample box containing the Teflon bags contributed up to a 10,000-fold value above the intrinsic levels of In and Ag.[4] Inasmuch as the samples were received under a dry nitrogen blanket, the contribution of surface nitrogen absorption was considered a possibility. Artifacts such as these from containers typify ultratrace contaminations that routinely occur, but unfortunately they are not always recognized.

In trace analysis the effects of contamination from laboratory air or furnishings, apparatus, containers, and reagents have become increasingly important, as the sensitivity of analytical methods has lowered detectable limits to the nanogram and picogram level. In addition, needs for accurate ultratrace analyses in air and water pollution, oceanography, biomedical research, clinical chemistry, geochemistry, electronics, and materials characterization have compelled analytical chemists to become more aware of the limitations imposed on their work by factors that were once judged too difficult or impossible to control. Clean-room techniques are now becoming broadly accepted and contamination-free sample handling techniques are rapidly developing. We have collected information on these advances to permit the reader to acquire a working knowledge of the principles and practices of ultrapurity techniques. The authors have also attempted to treat critically the various parameters that bias quantitative measurements at the submicrogram level.

Although contamination problems are well known to experienced trace analysts, some examples are included in the book primarily for the novice. Other details that an "academic" analytical chemist may consider to be trivial become vitally important to the real-world analyst. Wherever possible, information to demonstrate such effects has been provided throughout the book.

B. SCOPE

The absolute purity of chemicals and materials is assessed by quantitatively determining the individual impurities. This impurity content was related more than 25 years ago to terms that are now commonly used. The percentage ranges of impurities < 0.01, 0.01–1, and 1–100% were termed trace, minor, and major constituents. Today the 0.0001–0.01 and < 0.0001% ranges generally describe trace and ultratrace levels, respectively. In another classification $< 10^{-6}$, 10^{-2}–10^{-6}, and 10^{-2}–10^{+2}% ranges define ultratrace, trace, and major constituents.[5]

Analyses have also been characterized on the basis of the size of the analytical sample. The following classifications have been made: <0.001 g (submicro), 0.001–0.01 g (micro), 0.01–0.1 g (meso), and >0.1 g (macro).[6] In recent years ingenious methods have been developed for handling smaller and smaller samples. Indeed, a monograph on submicro analysis[7] has presented resourceful innovations for handling micrograms of materials. However the development of a specific determination at this reduced scale may take 1 or 2 man years of specialization.[8] Although such methods are too sophisticated for the average laboratory, they are particularly valuable when the amount of sample available for analysis is limited. When availability of sample is not a problem, an analysis can be based on a 5 or 10 g sample, particularly when the trace constituent must be isolated or preconcentrated.

Trace analysis includes the determination of contaminants and also the characterization of materials with respect to physical defects. Metallic, inorganic, and organic components are characterized, but most monographs on trace analysis have been concerned with inorganics and metals.[9-18] Measuring contaminants in these materials rather than defects is also the primary emphasis of this book.

Physical methods, the Hall effect,[19] resistivity,[13] and thermally stimulated currents,[20] have served as useful nonspecific techniques for indicating relative degrees of purity of solid state electronic materials. However precise and reliable quantitative methods for the characterization of specific impurities are also required. In most cases needs for specific determinations and identifications of impurities have been adequately fulfilled by refinements in instrumental methods of analysis that have lowered detection limits substantially in the past 20 years. Emission spectroscopy (ES), X-ray fluorescence (XF), atomic absorption spectroscopy (AAS), mass spectrometry (MS), and neutron activation analysis (NAA) now permit detection at the nanogram (10^{-9} g) level and frequently at the picogram (10^{-12} g) level, usually with aid of preconcentration techniques. Direct determination of 0.01–100 ng/g of various traces is possible with the high sensitivity of MS and NAA. When preconcentration is required, precipitation or coprecipitation, solvent extraction, ion exchange, or electrolysis can be used. Handling and containment of the sample during these procedures can be the most critical step in a determination at the nanogram level because of the possibility of contamination.

C. CONTROL OF PARAMETERS

The accuracy at which an impurity element can be measured at the ultratrace level is limited by systematic errors associated with each parameter shown in Figure 1. These parameters, which define the key phases of the overall analytical method, must be rigidly controlled during all aspects of the analytical procedure, from collection of the sample to the final detection and measurement of the property of the species undergoing measurement.

The knowledge and skill of the analyst are key factors. Constant evaluation and updating of professional competence are required for any

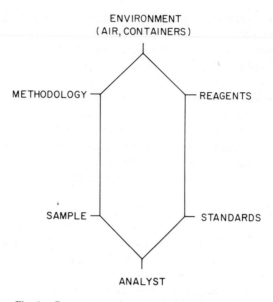

Fig. 1.   Parameters to be controlled in trace analysis.

analytical staff involved in ultratrace measurements. It is clear that the analysis can be no better than the analyst, since his judgments, decisions, and performance affect the validity of the results.

A perfect analysis of an improper sample is a futile exercise. Methods for the correct sampling of liquids, powders, and bulk solids are discussed in Chapter 6. Because of the inherent difficulties in the selection and preservation of an appropriate solid or liquid sample that contains nanogram per gram levels of such commonly occurring metals as aluminum, copper, lead, and iron, the following advice offered by Thiers is an excellent guideline: "unless the complete history of any sample is known with certainty, the analyst is well advised not to spend his time analyzing it."[21]

The analyst must select the appropriate methodology based on the physical and chemical nature of the sample as well as on the analytical information required. A comprehensive command of the chemistry involved, knowledge of interferences, matrix effects, and the detailed aspects of the analytical technique must be understood. An oversight of any kind can invalidate the analytical results. Detailed investigations must be performed to determine the validity of the methodology adopted for a given application.

Ultratrace analyses must be performed in controlled environments to protect the sample from artifacts contributed by laboratory air or unclean containers. Special procedures must be employed for the purification and storage of analytical reagents. Excessive amounts of reagents and prolonged contact of samples with containers at elevated temperatures contribute to the size and variability of the blank.[19] Control of these parameters is discussed in Chapters 3, 4, and 5, respectively.

In addition, certified reference standards are highly desirable. It has been estimated that certifies standards are unavailable for ~50% of all trace analyses. Although accuracy can be achieved in many of these cases with synthetic standards, the assurance of accuracy depends on the availability of certified standards. Chapter 2 discusses the preparation of synthetic standards and the availability of reference standards.

Rigid management of one parameter is not sufficient to obtain accuracy in trace measurements. Consequently results obtained in intercomparison studies by different laboratories frequently differ because all these parameters are not under complete control. Scrupulous efforts to eliminate deficiencies in all these factors must be undertaken. Such techniques are discussed in the remaining chapters of this book.

D. EXISTING NEEDS

1. Purity Determinations

Although no single quantitative definition of purity or ultrapurity has been universally adopted by the scientific community, several definitions have been advanced. In 1958 Melchior[22] characterized metals by R (Reinheitgrad, degree of purity) as follows: $R = -[\log (100 - w)]$ where w is the weight percent of the major element. For example, if a metal is 99.999% pure, $R = -\log 0.001 = 3$. In 1961 a British company suggested that the symbol N be used to denote nine in the following examples of purity designation: 5N indicates 99.999% and 4N5 indicates 99.995%.[23] The N value was assigned after totaling the impurity concentrations observed on a spectrographic plate and subtracting from 100. Later the letter Z was added to the purity designation to show that zone melting was the preparative method; 5N (Z) corresponds to a zone-refined product 99.999% pure.

Recently the prefix "m" has been introduced to indicate metallic impurities, whereas "t" designates purity based on total contaminants including oxygen, carbon, and nitrogen.[24] In this system designed for metals, "m5N5; t4N" indicates that an analysis by emission spectrography detected a total of 5 μg/g of impurities. By applying other analytical techniques, an additional 95 μg/g of trace contaminants was identified.

In 1965 the Soviet IREA (All-Union Scientific Research Institute of Chemical Reagents and Ultrapure Chemical Substances) proposed a system based on the number and concentration of specific impurities. A designation of 10-5 means that 10 trace elements have a total concentration of $1 \times 10^{-5}\%$. The second number in this system is the negative of the logarithm of the total concentration of 10 trace elements (expressed as percent).[25] The raw data for silica of 10-5 purity were as follows:

Element	Concentration (%)	035
Al	2×10^{-6}	
B	$<1 \times 10^{-7}$	
Ca	5×10^{-6}	
Fe	3×10^{-6}	
Mg	$<1 \times 10^{-6}$	
Na	$<5 \times 10^{-5}$	
P	2×10^{-7}	
Pb	$<5 \times 10^{-6}$	
Sn	$<6 \times 10^{-6}$	
Ti	$<4 \times 10^{-7}$	
Total	5×10^{-5}	

In this example the total of all impurities is considered to be less than $5 \times 10^{-5}\%$. Numbers above and including 5 are rounded off to 10. If the concentration of impurities totaled 6×10^{-5}, the value would be rounded off to 10×10^{-5} or $1 \times 10^{-4}\%$. The purity index would then be 4. Adoption of this logarithmic expression for the degree of purity would constitute a step forward in standardizing the literature. In any case the terms "ultrapure," "superpure," or "spectrographically pure" have no meaning unless accompanied by a detailed certificate listing the actual lot analysis, the method of analysis, and its detection limits, as well as results on similar materials with certified data for the traces of interest.

Determination of the purity of man-made materials and chemicals is one of the most important applications for methods of trace element analyses. In the early 1950s the explosion in solid state electronics research was catalyzed by earlier observations that electrical properties of semiconductors were directly related to the impurity content. This research actually generated the modern age of trace analysis when characterization of zone-refined germanium and silicon became a challenge in 1952.[26] Electronics, still a leader in the field, was among the first major disciplines in which the importance of ultratrace impurities in pure materials was well recognized.

The impurity limits for ultrapure materials vary with end use. For semiconductor applications, ultrapure metals with impurity tolerances in the low nanogram per gram (10^{-9}) range are required. On the other hand, ultrapure organics have been arbitrarily assigned a minimum purity of 99.95% corresponding to a maximum of 5×10^5 ng/g of impurities.[27]

Advances in new technologies have been critically dependent on the accurate characterization of ultrapure materials. For example, an increase in the capacity of telecommunication systems to handle the volume of messages expected in the 1980s depends on ultrapure components for optical waveguides that can transmit a light beam generated by lasers.[28] Since the target specifications for the maximum tolerable concentrations of transition elements in soda lime glass waveguides[29] (Table 1) are in the low nanogram range, extremely reliable analyses are required to evaluate materials. The specification for iron is probably the most difficult to attain because of the prevalence of iron in all atmospheres. To satisfy these transition metal requirements in the waveguide, the components for glass fabrication must be prepared and analyzed in a clean air environment.

The attainment of ultrapurity and the ability to assess it quantitatively have had significant impact on the development of several other technologies. The practical use of nuclear reactors was dependent on the fabrication of high purity graphite with ultralow concentrations of trace elements having high thermal neutron capture cross sections.[30] A third generation of high resolution radiation detectors constructed from intrinsic germanium was made possible by the recent preparation of pure germanium metal containing less than 0.001 ng/g (10^{-12}) of trace impurities.[31]

2. Characterization of Natural Systems

A second major application for trace analyses includes the quantitative scrutiny of naturally occurring systems (Table 2).[32] In the last 5 years

TABLE 1. Trace Element Specifications for
Soda Lime Glass Waveguides

Element	Maximum Concentration (ng/g)
Chromium	20
Cobalt	2
Copper	50
Iron	20
Manganese	100
Nickel	20
Vanadium	100

Source: Ref. 28. Copyright 1972, Bell Telephone Laboratories, Incorporated. Reprinted by permission, Editor, Bell Laboratories RECORD.

worldwide concern with the environment has catalyzed an enormous growth rate in the number of trace analyses performed on water, air, food, soil, plants, blood, and urine. The maximum levels for selenium, cadmium, lead, and hexavalent chromium in domestic water supplies are 10, 10, < 50, and 50 ng/g, respectively.[33] Tolerance limits for traces in other media are not easily established on the basis of scientific data.

The difficulty in relating systemic toxic action in the body to a specific pollutant and to its concentration is exemplified by the Minimata disease in Japan.[34] Medical records indicate that symptoms of the "strange" disease began to plague the Minimata area in 1953, yet not until September 1968 was methyl mercury found to be the causal agent. Again, since 1910 the strange "Itai-Itai" (ouch-ouch) disease had been endemic in Toyama on the Island of Honshu, but not until January 1967 was chronic cadmium poisoning identified as the basic cause.

Even though the Minimata and Itai-Itai diseases were confined to limited areas, they alerted many governments to the hazards of poisonous traces in water and air. The controversy about air quality standards for carbon monoxide, sulfur dioxide, and nitrogen oxides needs no repetition. For air quality measurements, traces must now be reported for potentially hazardous elements such as beryllium, mercury, arsenic, cadmium, vanadium, and manganese in particulate matter. The glass fiber filters used for collection, however, are commonly contaminated with ultratrace amounts of the element under investigation. Other extensively studied trace components in air now include polynuclear aromatic hydrocarbons that are suspected carcinogens in man. The details for quantitative analysis of these hydrocarbons have already been reported.[35, 36]

TABLE 2.   Applications for Trace Analysis

Application	Typical Matrix
High purity materials	
Electronics	Metals, inorganic dopants, alloys, single crystals, thin films, ceramics
Telecommunications	SiO_2, $SiO_2 + B_2O_3$, $SiO_2 + TiO_2$, $SiO_2 + Na_2CO_3 + CaCO_3$, GaAs, GaP
Standards and reagents	Acids, bases, a variety of inorganic and organic chemicals
Natural systems	
Stratosphere	Lunar samples, air
Atmosphere	Air, particulates
Hydrosphere	Seawater, potable water, industrial and ground water
Lithosphere	Minerals, soils, agricultural commodities
Biosphere	Blood, urine, tissues

Determinations of trace and ultratrace concentrations of metals in the human body are now assuming increasing importance in assessing exposure to toxic substances and relating enhanced concentrations of these elements to arteriosclerosis, diabetes, and other diseases.[37, 38] At these levels special handling techniques must be superimposed on the parameters usually connected with methodology in clinical chemical laboratories. Some hospitals send samples to a commercial or special regional laboratory for ultratrace analysis, but contamination during packaging of these samples in a conventional laboratory and subsequent shipment may well vitiate the results.[39]

E. CHALLENGES

Accuracy in trace analysis is difficult to obtain. In reviewing chemical oceanographic data, Carpenter commented that "for the trace constituents including many of the heavy metals, inaccuracy by current techniques of analysis has produced a nearly useless body of data."[40] This statement might well apply to trace element data reported in many of the disciplines listed in Table 2. Eliminating the problems that contribute to these inaccuracies currently constitutes the major challenge to trace analysts. The inability to control the blank at levels insignificant in comparison with the constituent being determined severely restricts the accuracy and precision of ultratrace measurements. Contamination from particulates in the air, impurities in reagents, and leaching of trace elements from containers are primary causative factors of high blanks. Hazards from less conspicuous sources, including instrumental noise interpreted as a signal from a component of the sample, must also be considered.[41] Most trace element determinations suffer from positive (airborne contamination or leaching from containers), negative (adsorption or volatilization), or pseudo (method of sampling) forms of contamination described by Thiers.[42] Other problems in dissolving, separating, and preconcentrating ultratrace elements have been reported by Tölg.[43]

Obviously quantitative determinations of the influence of trace elements in the environment, in biological functions, in determining mechanisms by which heavy metals induce toxicity, and in the physics and chemistry of semiconductors will yield meaningless results if significant environmental or analyst bias is operative during the experiment. To meet the present and future challenges in trace characterizations, therefore, analytical and materials scientists, nutritionists, biochemists, forensic scientists, environmentalists, oceanographers, geochemists, solid state researchers, and others must practice the art of "ultrapurity"—an endeavor that has advanced significantly in recent years.

This book describes state-of-the-art techniques and methods for providing pure atmospheres, for preparing and storing ultrapure reagent chemicals, and for performing routine analytical procedures under ultraclean conditions. Recognition and elimination of the contamination sources that cause mistakes in trace analysis are also delineated, and a critical evaluation of commercially available apparatus or equipment is included.

REFERENCES

1. M. Tatsumoto and J. N. Rosholt, *Science, 167,* 461 (1970).

2. P. A. Baedecker and J. T. Wasson, *Science, 167,* 503 (1970).

3. G. H. Morrison, J. T. Gerard, A. T. Kashuba, E. V. Gangadharam, A. M. Rothenberg, N. M. Potter, and G. B. Miller, *Science, 167,* 505 (1970).

4. R. R. Keays, R. Ganapathy, J. C. Laul, E. Anders, G. F. Herzog, and P. M. Jeffery, *Science, 167,* 490 (1970).

5. K. Heydorn, Seventh Materials Research Symposium, National Bureau of Standards, Gaithersburg, Md., October 7-11, 1974.

6. E. B. Sandell, *Colorimetric Determination of Traces of Metals,* 3rd ed., Wiley-Interscience, New York, 1959.

7. G. Tölg, *Ultramicro Elemental Analysis,* Wiley-Interscience, New York, 1970.

8. G. Tölg, *Talanta,* **21,** 327 (1974).

9. I. P. Alimarin, ed., *Analysis of High-Purity Materials,* J. Schmorak, transl., Israel Program for Scientific Translations, Jerusalem, 1968.

10. I. P. Alimarin and M. N. Petrikova, *Inorganic Microanalysis,* M. G. Hall, transl., Pergamon Press, London, 1964.

11. J. P. Cali, ed., *Trace Analysis of Semi-Conductor Materials,* Pergamon Press, London, 1963.

12. O. G. Koch and G. A. Koch-Dedic, *Handbuch der Spureanalyse,* 2nd ed., Springer-Verlag, Berlin, 1974.

13. I. Korenman, *Analytical Chemistry of Low Concentrations,* J. Schmorak, transl., Israel Program for Scientific Translations, Jerusalem, 1968.

14. W. W. Meinke and B. F. Scribner, eds., *Trace Characterization (Chemical and Physical),* National Bureau of Standards Monograph 100, Government Printing Office, Washington, D.C., 1970.

15. L. Meites, ed., *Handbook of Analytical Chemistry,* McGraw-Hill, New York, 1963.

16. G. H. Morrison, ed., *Trace Analysis: Physical Methods,* Interscience, New York, 1965.

17. M. Pinta, *Detection and Determination of Trace Elements,* Humphrey, Ann Arbor, Mich., 1969.

18. J. H. Yoe and H. J. Koch, Jr., eds., *Trace Analysis,* Wiley, New York, 1957.

19. T. J. Murphy, Seventh Materials Research Symposium, National Bureau of Standards, Gaithersburg, Md., October, 7–11, 1974.

20. C. Kittel, *Introduction to Solid State Physics,* Wiley, New York, 1965.

21. R. E. Thiers, in *Methods of Biochemical Analysis,* D. Glick, ed., Vol. 5, Interscience, New York, 1957, p. 274.

22. P. Melchior, *Metall,* **12,** 822 (1958).

23. Light and Co., England, *Catalog of Ultrapure Elements,* 1961.

24. Alfa Products, Ventron Corp., Beverly, Mass., 1972–1973 Catalog, p. 147.

25. B. D. Stepin, I. G. Gorshteyn, G. Z. Blyum, G. M. Kurdycmov, and I. P. Ogloblina, *Methods of Producing Superpure Inorganic Substances,* Leningrad, 1969. (Reproduced by National Technical Information Service, Springfield, Va. 22151.)

26. W. G. Pfann, *Trans. AIME,* **194,** 747 (1952).

27. W. Wilcox. R. Friedenberg, and N. Back, *Chem. Rev.,* **64,** 194 (1964).

28. A. D. Pearson and W. G. French, *Bell Laboratories Record,* **103,** April 1972. Bell Telephone Laboratories, Inc., Murray Hill, N.J.

29. J. S. Cook, *Sci. Am.,* **229,** (5), 30 (1973).

30. W. C. Riley, ASTM Technical Publ. **276,** 1960, pp. 324–355.

31. *Industrial Research,* December 1970, p. 29.

32. W. W. Meinke and J. K. Taylor, eds., *Analytical Chemistry: Key to Progress on National Problems,* National Bureau of Standards Special Publication 351, August 1972.

33. Report of the Committee on Water Quality Criteria, Federal Water Pollution Control Administration, U.S. Department of the Interior, Government Printing Office, Washington, D.C. 20402, 1968.

34. *Pollution Related Diseases and Relief Measures in Japan,* Environment Agency, Japan, United Nations Conference on the Human Environment, May 1972.

35. E. Sawicki, *Proc. Arch. Environ. Health,* **14,** 46 (1967).

36. D. Hoffman and E. L. Wynder, in *Air Pollution,* A. C. Stern, ed., Vol. 2, Academic Press; New York, 1968.

37. D. R. Williams, *Chem. Rev.,* **72** (3), 203 (1972).

38. H. A. Schroeder and A. P. Nason, *Clin. Chem.,* **17** (6), 461 (1971).

39. M. Zief and F. W. Michelotti, *Clin. Chem.,* **17,** 833 (1971).

40. R. H. Bube, *J. Chem. Phys.,* **23,** 18 (1955).

41. Ref. 20, p. 393.

42. D. E. Robertson, in *Ultrapurity: Methods and Techniques,* M. Zief and R. M. Speights, eds., Dekker, New York, 1972, p. 208.

43. G. Tölg, *Talanta,* **19,** 1489 (1972).

CHAPTER 2

BASIC ASPECTS OF QUANTITATIVE ULTRATRACE ANALYSIS

Many problems arise and several pitfalls exist during the quantitative determination of ultratrace elements. Eliminating the effects of contamination, a major concern, can be partially circumvented by direct analyses based on highly sensitive, nondestructive, instrumental methods. Although this approach minimizes the introduction of contaminants by reducing the amount of sample handling and pretreatment, sophisticated instrumental methods are not free from contamination problems. For example, fluorescent or electroactive organic traces in water interfere with fluorometric and electrochemical measurements. Residual hydrocarbons in ultravacuum systems, tenacious absorption of atmospheric gases on sample surfaces, and memory effects due to reversible adsorption and desorption of components on walls of instrumental system are only a few of the events that limit the analytical application of several sophisticated instruments. In addition to the problems that result from contamination, fluctuations in the level of instrumental noise, nonavailability of standards, matrix effects, problems in measurement, and difficulties in recovery of traces in dilute solutions are common.

Chemical methods, in spite of the time-consuming effort involved, have proven to be superior to most semiautomated, multielement, instrumental techniques when accurate and precise measurements for a few elements are important. Whether speed or high accuracy is more important is determined by the need for the analytical information. Since a laboratory handling a diversity of analytical problems must accommodate both cases, a well-balanced mixture of fast survey instrumental techniques and highly accurate chemical methods must be maintained. Methods for simultaneous determination of many elements or those used for survey identifications are listed in Table 1. Some single element and compound specific techniques are given in Table 2. Since the instrumental methods in Table 1 can be rendered inapplicable by matrix effects or insufficient resolution, the latter group (Table 2), which allows sample modification before analysis, often presents suitable alternatives for coping with these difficulties. Indeed, it has been shown that established classical methods can be used to determine nanogram quantities of traces accurately.[1, 2]

A renaissance in the use of microanalytical techniques might well prove

TABLE 1. Techniques for Simultaneous
Instrumental Determinations or Identifications of
Trace Elements

Method
Emission spectroscopy[a]
X-Ray fluorescence[b]
Mass spectroscopy (spark source)
Activation analysis (particle and neutron)
Ion and electron microprobe
Auger spectroscopy
Rutherford backscattering

[a] Includes all forms of excitation.
[b] Energy dispersive.

to be essential in obtaining the ultimate sensitivity with current analytical technology. Traces, concentrated properly into microliter volumes, can be determined at concentrations orders of magnitude lower than presently possible. Incorporating microchemical methods in the training of graduate analytical chemists would stimulate the development of the quantitative perception required to perform ultratrace determinations. Skills in manipulating and recovering microgram amounts of materials, accurate measurement of microliter volumes, and techniques for weighing microgram samples are

TABLE 2. Techniques for Single Element Determinations

Method
Volumetric analysis
Ion-specific electrodes
Coulometric titration[a]
Spectrophotometry
Spectrophotofluorometry
X-Ray fluorescence with crystal optics[a]
Kinetic measurements[a]
Atomic absorption spectroscopy[b] or atomic fluorescence spectroscopy
RF induction coupled atomic emission
Electron-capture gas-liquid chromatography[a]
Stable isotope dilution[a]
Radioisotope dilution
Nuclear track counting
Polarography[a]

[a] Potential for sequential element determination.
[b] Flame and carbon rod furnace.

necessary regardless of the analysts' specialty. When it is necessary for high accuracy to be the most important criterion for measurement, analytical chemists in the more industrialized countries should reexamine the utility of tailoring microspectrophotometric, -fluorometric, -titrimetric, and -radiometric methods for ultratrace determinations.

Although analyses are susceptible to considerable bias at the ultratrace level, satisfactory results can be obtained if unusual attention is given to all conceivable variables. For quantitative work the most damaging factors include sample loss, inadequate trace standards, inaccurate measurements of volume and weights, errors in calibration of instrument parameters, and other systematic errors due to the blank. Additionally, improper handling will compromise the integrity of the sample, convey a meaningless qualitative analysis, and yield a quantitative result that does not truly represent the original constituency of the sample.

I. LITERATURE ON ULTRATRACE ANALYSIS

Often the obstacles and pitfalls encountered during the perfection of a reliable quantitative method are not thoroughly discussed in analytical papers. Analytical chemists frequently assume that procedures for ultratrace analysis can contain too much detail concerning the execution of the method. Indeed, a cognizance of the "art of trace analysis"—recognition of the important details that require unusual attention in perfecting a highly accurate method—is largely acquired through experience with the technique. If the pool of knowledge about heretofore unsuspected factors that affect analytical results at the trace level is to be augmented, however, it is imperative to report sufficient detail and to discuss factors critical to the successful application of the method. Full description of all pertinent details facilitate application of a method by others, whereas fragmentary reports of techniques for ultratrace analysis can seldom be adopted even by skilled analytical chemists.

Methodology at the ultratrace level may be a unique combination of events that must be precisely followed and exactly reproduced by analysts desiring to apply a method reported in the literature. Thus in ultratrace work it is not sufficient to report that a difficult quantitative measurement was successfully accomplished with good accuracy and precision. Unless the author has recorded the procedure in every detail, even he may encounter problems in checking his own work. Trace analytical chemists must therefore become proficient observers, record keepers, and reporters of all details pertaining to ultratrace measurements.

II. THE BLANK

In ultratrace analysis the size and fluctuation of the blank must be reduced as much as possible. During an actual analysis the lower limit of detection for a given element is seldom restricted by the sensitivity of the method but is more often determined by the size and variability of the blank. A qualitative picture of this situation is depicted in Figure 1. For an analytical instrument that has sufficient sensitivity, a signal that preferably has a linear relationship with concentration can be measured with high precision. Usually for quantities of the element above 1 μg, relative standard deviations of $\pm 2.0\%$ or better can be obtained routinely. For smaller quantities of the element, the range of values representing replicate measurements increases considerably because of positive contributions from the blank and losses of the element.

When the concentration of the element in the sample becomes negligible in comparison to the blank, a variable signal due solely to the blank value is measured (curve A). A least square line drawn through data in the appropriate region gives the magnitude of the blank. A blank of 10 ng with a variability of ± 5 ng is shown. In practice this low blank usually can be

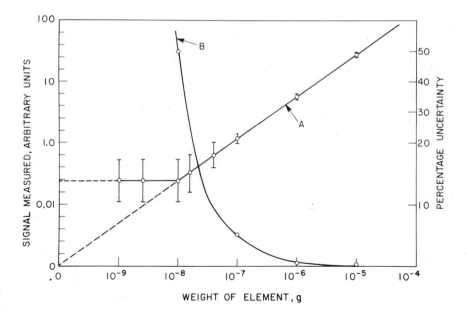

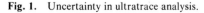

Fig. 1. Uncertainty in ultratrace analysis.

exceeded by an order of magnitude even in very careful work. For a blank of this size, the uncertainty involved in the measurement of trace elements is considerable as demonstrated previously by Murphy,[3] and curve B of Figure 1 was constructed following his representation. It is clear from this graph that the accuracy of the measurement is inversely related to the variability of the blank. Furthermore as the amount of the element in the sample approaches the size of the blank, subtraction of the portion of the signal due to the blank value becomes a highly inaccurate exercise of obtaining small differences between two values.

The relationship between the detection limit, the blank, and its variability is given by

$$\text{D.L.} = \bar{x} + k\sigma_{\text{bl}}$$

where D.L. indicates the limit of detection, $\bar{x}$ is the average value of the blank, σ_{bl} is the standard deviation of the blank, and the factor k has been assigned a value of 3.[4] This statistical definition shows that the threshold below which ultratrace determinations are impossible is mainly established by the magnitude and reproducibility of the blank value. Consequently the optimum sensitivities for many analytical methods are well below the levels attainable in practical analyses of real samples (see Table 3).[5]

Blank determinations can show higher values of the element in question than might be found in the actual sample. If the blank is highly reproducible, it can be subtracted from the analytical result to give a corrected value. The precision of the blank is more important than its size. Nevertheless, a blank level less than one-tenth the quantity of the element undergoing measurement is usually required for accurate ultratrace determinations.

When the blank value approaches the detection limit of the analytical technique, measuring the blank is difficult indeed. In this case contribution to the blank from reagents can be measured by using 10 times more reagent for the blank determination than is used for the sample. Thus the blank can be measured in a range at which the magnitude may be determined with reasonable accuracy, and meaningful corrections can then be made if the reagent blank is the major component of the overall blank.

In ultratrace analysis it is essential that blank determinations be performed in the same way and at the same time as the sample, if reliable corrections for the blank are to be made. An example of reliable determinations of the blank during analysis for trace elements is cited in Table 4. The data were obtained for repetitive determinations of blanks during vapor phase destruction (Chapter 6) of high purity silicon dioxide and subsequent analysis by the "Coprex" microdot X-ray fluorescence method (Chapter 7). Even without conversion of the X-ray counts to nanograms of elements, it is

TABLE 3. Ideal Detection Limits for Analytical Methods

Method	Limit of Detection[a] (g)	Remarks
Titrimetric methods	10^{-9}	Electrometric indication
Spectrophotometry	10^{-10}	Capillary cells
Fluorimetry	10^{-11}	Capillary cells
X-Ray fluorescence	10^{-9}	Curved crystals
Electron probe micro analysis	10^{-14}	Calculated
Polarographic methods	10^{-10}	Stripping methods
Gas chromatography	10^{-12}	Electron-capture or flame-ionization detection
Atomic absorption spectroscopy	10^{-13}	Nonflame technique
Emission spectroscopy	10^{-10}	Solution methods
Catalytic methods	10^{-12}	
Isotope dilution and radiochemical methods	10^{-12}	Substoichiometric extraction
Neutron activation analysis	10^{-11b}	Large cross sections, short half-lives
Mass spectroscopy	10^{-16}	Calculated

[a] Favorable limit of detection for a particularly easily determined element. These values can only be very approximate estimates for reasonably favorable conditions and may be very readily influenced by many experimental factors.
[b] For neutron flux of 10^{14} neutrons/$(cm^2)(sec)$.
Source: Ref. 4.

clear from the data that subtraction of data in columns 3 and 4 show the presence of nickel, iron, and chromium in the sample.

The high precision of these blank measurements resulted from precise control of the environment in which the sample was decomposed in a Teflon chamber. It is of paramount importance that contamination from the environment and from containers be kept as small and as reproducible as possible during the blank and sample measurement. For example, the large deviation of the mean value for iron in the sample was not due to systematic errors. Since the reproducibility of blanks determined simultaneously was acceptable, the highly variable results for the sample were caused by inhomogeneous distribution of iron in the original sample.

The identification of all factors contributing to the blank is a difficult task. The only good rule to follow when investigating the cause of the blank is that until proved otherwise, everything and all procedures are assumed to contribute to blank size.

TABLE 4.   Determination of Blanks During Vapor Phase
Destruction of Silicon Dioxide[a]

| | X-Ray Counts (measured per second) | | | |
Cation	Blank	100 ng of Cation	Sample + 100 ng of Cation[b]	Net Counts
Cu	598 ±4	965 ±37	947 ±21	N.D.
Ni	170 ±2	631 ±5	684 ±2	53
Co	80 ±2	371 ±16	346 ±20	N.D.
Fe	126 ±15	489 ±25	931 ±315	442
Mn	46 ±1	232 ±9	254 ±25	N.D.
Cr	79 ±0	136 ±10	355 ±47	219

[a] Mean of four determinations by J. E. Kessler, Bell Laboratories,
Murray Hill, N.J.
[b] Corrected for blank.

Several approaches have been used successfully to diminish the effect of
the blank. By using devices that simultaneously detect signals from a
sample and reference cell containing the blank, an output signal equivalent
to the difference of the two detectors automatically compensates for the
blank. Techniques based on relative rates of reactions also have an internal
capability of compensating for the blank. The effects of instrumental noise
as a contributor to the blank have been lowered significantly by internal
standardization. Daily presetting of instrument signals with reference
standards also corrects for slow instrumental drift. Standard addition techniques are highly successful, as well.

Truly nondestructive instrumental methods are obviously susceptible to
smaller blanks than are techniques requiring sample preparation. However
other difficulties involving reference standards and matrix effects may
constitute a greater analytical problem than the blank itself. A significant
contribution toward eliminating the blank problem can be provided by a
method that stimulates characteristic atomic, chemical, or nuclear
properties in atoms of interest in the sample, to distinguish them from the
same element that may be introduced subsequently during physical or
chemical treatment. Thus radioactivation is an excellent example of this
technique.

III.   ACCURACY AND PRECISION

The accuracy required for an analytical measurement is determined by
the intended use of the data. The quantitative result need not be any more

accurate than necessary to solve the immediate analytical problem. However an expert in trace analysis must develop the skills needed to achieve the highest accuracy possible with the technique he selects as a specialty. Once this expertise is acquired, it is easy to relax stringent procedures for high accuracy quantitative measurements, to ensure that semiquantitative or qualitative data are obtained. On the other hand, once rough skills and techniques become a habit, developing the quantitative instincts that are necessary to perform high accuracy trace measurements is extremely difficult. It usually takes 6 months to several years of experience to acquire skills necessary to perform high accuracy ultratrace analyses. The degree of accuracy achieved should be used as the true measure of the analyst's success.

Accuracy (derived from the Latin *accuratus*) is defined as a state of being free from mistakes. When the mean value $\bar{x}$ of a number of such individual determinations x_i is computed, an accurate value is obtained. The degree of accuracy of the measurement is defined by the closeness of the result x_i to the true value μ. Precision measures the reproducibility of a series of results without regard to their agreement with the true value. The simplest measurement of precision, average deviation d, expresses the average of the deviation of the individual result x_i from the mean $\bar{x}$ without regard to sign:

$$d = \frac{1}{n} \sum_{i=1}^{n} (x_i - \bar{x})$$

Standard deviation s is more useful in trace analysis because it more adequately reflects the presence of widely differing values:

$$s = \left[\frac{1}{n-1} \sum_{i=1}^{n} (x_i - \bar{x})^2 \right]^{1/2}$$

Thus the relative standard deviation $s/\bar{x}$ (100) is generally accepted for expressing precision or reproducibility.

Recently a practical and easily comprehended example of accuracy and precision was presented at a symposium.[6] The standard can be visualized as the bulls-eye of a target. Accurate shooting of the marksman is assessed by the distance of the bullet hole from the bulls-eye. If the marksman clusters several shots in a narrow area some distance from the bulls-eye, his results are reproducible. He then exhibits good precision but poor accuracy.

Each step of an analytical procedure has a corresponding precision that contributes to the overall variability of the analytical result. Testing the error and precision of each step allows the sources of significant contribution to the overall error of the method to be determined. Once the sources of errors are identified, the procedure can be altered to yield better precision and improved accuracy. Such detection of systematic errors by the analysis of variance has been successfully used in activation analyses.[7]

To establish the accuracy of an analytical method for the analysis of an unknown sample, different and independent techniques must be used. One of the methods should be a reference technique, that is, one of demonstrated accuracy.[6] Parallel analyses of standard reference materials (defined as accurately characterized materials in terms of certain properties[6]) must also be performed when such standards are available.

Periodic checks are necessary during routine analyses to detect systematic errors. Within a given laboratory this is accomplished by submitting reference standards to be analyzed with the different techniques available. The analysts should know that periodic unannounced accuracy tests will be conducted. They should not be informed that a test is being made, and the correct analytical value should not be disclosed until all participants have reported their data. Similar procedures should be adopted for round-robin analyses by different analytical laboratories. The trace element under investigation should be accurately determined in a standard sample by reference methods. Each participant in the study should then be given a portion of the standard along with the sample to be characterized. Comparison of the two results facilitates the detection of systematic errors among the participants. Results from round-robin analyses without simultaneous measurements on a reference sample are usually very difficult to interpret.

IV. STANDARDS

A. SYNTHETIC

Frequently the calibration of analytical instruments and testing of trace methods can be based only on results obtained by using synthetic standards, since suitable primary reference standards certified for trace element content are scarce. When faced with preparing the standard the analyst must consider that (1) the standard matrix should closely approximate that of the sample, (2) the concentration of elements in the artificial standard should be of the same order of magnitude as those expected to occur in the sample, and (3) the concentration of the desired element in the standard must be known to a high degree of accuracy. The first requirement has often been impossible to meet. Thus standards containing the element of interest, but not all matrix constituents, have been routinely prepared. Obviously comparison of results from such standards with real samples will not give valid information if matrix effects and interferences are present.

1. Solids

Solid standards are particularly difficult to prepare. Doping a bulk solid with a given trace is rarely better than $\pm70\%$ of the intended value and homogeneous distribution of the dopant in the solid matrix is extremely difficult to attain. Additionally, unpredictable contributions by way of contamination or losses require quantitative analysis by two or more techniques to determine the trace elemental content of the synthetic standard. This certification of the standard is not a trivial matter.

Melting solids and mixing dopants by stirring has met with some success. However radial or lateral segregation of the dopant is difficult to eliminate even in reasonably low melting alloys. Another method, developed by the author to produce a homogeneously doped SiO_2 powder, involves mixing tetraethylorthosilicate and an acidic solution doped with the desired traces and subsequently reacting the mixture chemically to generate the matrix. Upon hydrolysis of the mixture during vigorous stirring, a homogeneous gel is formed. After drying, the gel is crushed to give a homogeneous powdered material containing the desired trace impurities. Although the homogeneity of the sample has been shown to be acceptable by preparing and counting test samples doped with radioisotopes, certification of the trace element content is still under investigation.

Recently ion implantation of a metallic matrix with trace elements was used to prepare standards for flameless atomic absorption spectrometry.[8] The method shows promise for preparing synthetic solid matrices with a single dopant for use as comparison standards. Although the method has been proposed for preparing primary trace standards and has potential, these possibilities are still quite optimistic.[9]

2. Liquids

Analyzing dissolved samples and using standard solutions serves to eliminate some of the problems encountered during the determination of traces in solids. However this approach can introduce interferences—for example, traces of complexing agents, solvent, and interelement effects. Also the stability of the standard solution must be considered, and several general rules must be followed when preparing standard solutions.

1. Prepare solutions immediately before use by diluting stock solutions containing at least 1 $\mu g/ml$ of the desired element.

2. Use plastic or quartz measuring utensils for the preparation of a given standard at a specified concentration. Avoid cross-contamination of all utensils.

3. Pre-equilibrate the walls of a freshly cleaned vessel with a portion of a solution containing the same concentration of the trace elements as the standard.

4. The standard solution should contain the desired element at a concentration $\geq 10^2$ times any blank resulting from the procedure used for preparing the solution.

The dissolution of etched, preweighed samples of pure wire or pellets in enough pure acid to give a final hydrogen ion concentration of 0.1 M is a popular method for preparing standard solutions. The purity of the sample must be carefully considered. The uncertainty factor for the percentage purity quotation, which is often given for wires, pellets, and metal products, can easily be ≥ 100. For example, metal or inorganic products quoted as 99.999% pure might well be 99.9, since the percentage purity is usually based on emission spectrographic analysis for cationic traces only. Oxides, halides, and gases are usually neglected. Nevertheless appropriate procedures permit the use of these materials for preparing 1 μg/ml standard solutions that have negligible amounts of other trace elements. Also pre-etched and dried metal can be accurately weighed. An assessment by use of radiotracers of losses occurring during the dissolution and transfer of the sample to a volumetric vessel allows the final concentration of the standard solution to be corrected.

Preparing standard solutions by dissolving ultrapure water-soluble salts in acidified, deionized, distilled water is subject to fewer problems than the treatment of metals with acids. Available salts, however, are frequently nonstoichiometric or consist of a mixture of hydrates. In these cases solutions of accurately known concentration cannot be prepared by dissolving weighed samples.

Anhydrous inorganic fluorides or stoichiometric hydrates have been prepared for use as activation analysis standards at this laboratory by zone melting of highly pure starting materials.[10] Microwave heating has been found to be very useful for dehydrating water-soluble salts for subsequent use as standards. The universal problem of particulate matter in powdered reagents[11] is solved by filtering the prepared solution through a purified 0.2 μm membrane filter. Appropriate volumes of the filtered solution can then be analyzed to determine the concentration of the element of interest.

B.   HIGH PURITY STANDARDS (MAJOR ELEMENT OR COMPOUND STANDARDS)

The purity of a major element or compound standard is seldom a factor limiting the accuracy of analytical results. Calcium carbonate, sodium chloride, potassium chloride, potassium dichromate, and boric acid are supplied commercially with detailed certificates of analysis. Standardization of these materials by high precision and high accuracy methods such as coulometry are necessary for establishing the actual assay value within 1 part in 10,000 or better. Frequently errors in weighing of high purity reagents are not taken into account during assay determination. For example, failure to correct for buoyancy of air can lead to errors of 0.05% or more in the assay determination.

Calcium carbonate is suitable as an acid-base, chelometric, or calcium standard. High purity ammonium and potassium monobasic phosphate, potassium bromide, and anhydrous magnesium acetate fully characterized by direct assay as well as cation and anion content are receiving attention as standards in the clinical laboratory.

The evolution of a suitable serum magnesium standard typifies problems usually encountered in selecting an appropriate standard. Ideal criteria in this case include excellent stoichiometry, relatively high equivalent weight, good solubility in water, and little gain in weight on contact with moist air. Various high purity materials that have been used as magnesium standards are listed in Table 5. Magnesium metal must be acid-washed, dried, then dissolved in acid; the oxide picks up water rapidly, and magnesium acetate dihydrate contains some basic acetate that destroys exact stoichiometry. The anhydrous acetate is an improvement in that it picks up water less rapidly than the oxide. The rate of moisture pickup by magnesium gluconate, however, is significantly less than that for the acetate or oxide.[12] In addition Table 5 indicates that weighing precautions are less crucial for magnesium gluconate. This example points out the contributions offered by organic derivatives of metals.

Analysts can prepare high purity materials for use as standards when needed. However it is usually more economical to purchase certified standards when these are available. For example, if an analyst decides to purify

TABLE 5. Some Magnesium Standards

Material	Formula	Formula Weight	Weighing Factor[a]
Mg metal	Mg	24.305	1
Mg oxide	MgO	40.304	0.603
Mg acetate, anhydrous	$Mg(OCOCH_3)_2$	142.397	0.1707
Mg acetate, tetrahydrate	$Mg(OCOCH_3)_2 \cdot 4H_2O$	214.584	0.1133
Mg gluconate dihydrate	$Mg(OCOC_5H_{11}O_5)_2 \cdot 2H_2O$	450.636	0.054

[a] $FW_{Mg}/FW_{compound}$.

sodium chloride, the cost of analyzing the final product alone could make the preparation prohibitive. The full cost of certification for sodium chloride incurred by the National Bureau of Standards (NBS) is several hundred dollars, but this fixed cost is manageable when distributed over a batch of 100 lb or more.

Aliphatic and aromatic compounds that melt without decomposition have been isolated as IUPAC grade C primary standards.[13] Ultrapurification by zone refining[14] after prepurification by such classical techniques as fractional distillation or recrystallization has provided such simple compounds as benzene, benzoic acid, acetanilide, and *p*-dibromobenzene in 99.99% purity. Polynuclear hydrocarbons (anthracene, naphthalene, phenanthrene, and pyrene) are also commercially available in 99.99% purity. For the chemist who wishes to prepare his own ultrapure organic standards, a fully automatic zone refiner is an invaluable aid.[15]

C.   STANDARDS FOR TRACE ANALYSIS

Recently an increasing number of certified standards for trace analysis has become available. Production and certification of such standards, however, are difficult and expensive operations as demonstrated by projects carried out at the NBS. The most ambitious project yet undertaken for producing and certifying standard materials was the preparation of an inorganic matrix consisting of 72% SiO_2, 12% CaO, 14% Na_2O, and 2% Al_2O_3, containing 61 trace elements.[16] More than 11,000 man-hours of analytical time went into the project, and an equal or greater expenditure is expected before most of the elements are certified. At least two or more results run independently by different scientists had to be in agreement to obtain certification.[16, 17] Table 6 lists trace elements in the glass and gives the status of certification.

NBS Standard Reference Materials, SRMs 610 to 619, will be valuable in establishing the accuracy of methods used in geology and geochemistry, for quality control processing of specialty glasses, and for testing instrumental trace methods. Their practical utility would be much greater, however, if groups or families of elements had been present in the matrix rather than the entire block of 61 trace elements. On the other hand, these SRMs provide excellent samples for establishing schemes for multielement analysis following group separations.[18, 19]

Several high purity metal standards for trace analysis are available from the NBS.[20-22] High purity platinum and doped platinum, SRMs 680 and 681, are available with preliminary certification for Ag, Au, Cu, Fe, Ir, Mg, Ni, O, Pb, Rh, and Zr. SRM 685, a high purity gold wire or rod containing less than 1 ng/g of Cu, In, Fe, and Ag, can be used in applications for cali-

TABLE 6.   Trace Elements in SRMs 610–619 Glass Standards

(Doped at 0.02, 1.0, 50, and 500 $\mu g/g$)

Antimony[a]	Fluorine	Molybdenum	Tantalum
Arsenic	Gadolinium[a]	Neodymium[a]	Tellurium
Barium[a]	Gallium[a]	Nickel[a]	Terbium
Beryllium	Germanium	Niobium	Thallium[b]
Bismuth	Gold[a]	Phosphorus	Thorium[c]
Boron[a]	Hafnium	Potassium[a]	Thulium
Cadmium	Holmium	Praseodymium	Tin
Cerium[a]	Indium	Rhenium	Titanium[c]
Cesium	Iron[b]	Rubidium[c]	Tungsten
Chlorine	Lanthanum[a]	Samarium[a]	Uranium[c]
Chromium	Lead[c]	Scandium[a]	Vanadium
Cobalt[a]	Lithium	Selenium	Ytterbium[a]
Copper[b]	Lutetium	Silver[b]	Yttrium
Dysprosium[a]	Magnesium	Strontium[a]	Zinc[a]
Erbium[a]	Manganese[b]	Sulfur	Zirconium
Europium[a]			

[a] Element has been studied, but there exist insufficient data for certification.

[b] Element concentration certified at one or more levels.

[c] Element concentration certified at all four levels.

brating instruments. High and intermediate purity zinc metal standards, SRMs 682 and 728, are certified for Pb, Cu, Fe, Si, and Cd.

Producing and partially certifying SRM 1577, the first available animal tissue standard, was an involved operation.[23] Following homogeneity testing on random samples, the SRM was certified for the following elements: Cd, Cu, Fe, Pb, Mn, Hg, N, K, Se, Na, Pb, and Zn. Information on As, Ca, Cl, Co, Mg, Mo, Ag, Sr, Tl, and U is also available.

The first of a series of botanical standards, SRM 1571 (orchard leaves), is available with certification for Ca, K, Fe, Na, Cu, and Ni.[24] Information on the trace content for N, Mg, P, As, Bi, B, Cr, Co, Fe, Mn, Se, U, and Zr is given. This standard provides a reference for the determination of trace elements in leaves of plants and in other biological materials. Producing SRM 1571 required well over 6 man-years of scientific effort at a cost of over $380,000 for analytical work.[25] Reference standards of fly ash and homogenized coal were recently introduced. Standards for mercury in water; lead in gasoline, citrus and tomato leaves, pine needles, and alfalfa; and NO_2 permeation tubes are in progress. The following are under consideration: air, oyster meat, spinach, grain, industrial hygiene matrices, steel, high temperature alloys, and rare earth glasses.

Since the National Bureau of Standards started its Standard Reference Materials program in 1906 at the request of the steel industry,[26] more than 800 SRMs have been made available in 70 different industrial and scientific categories.[27] These fields include chemicals, polymers, gases, biological materials, and radioisotopes. All types of physical and chemical standards for instrumental methods of analysis are also covered. Along with NBS, the American Society for Testing and Materials (ASTM), founded in 1898, has evolved as one of the world's largest sources of standards and test methods. In recent years 33 volumes of procedures for testing have been revised annually to reflect recent advances.[28] Volume 32 is devoted in large part to chemical analysis. In 1974 the Book of Standards was restructured, and revisions will expand the Annual Book to 47 volumes.[29]

Detailed procedures for preparing and storing standard solutions made from SRMs may be necessary in the future. A compilation of methods for pesticide analysis is an excellent example of quality control for standards.[30] Concentrated stock standards containing 200 μg/ml of the common chlorinated pesticides are prepared in pesticide quality (distilled in glass) hexane, benzene, or isooctane (2,2,4-trimethylpentane). Isooctane is the preferred solvent because its higher boiling point reduces losses from evaporation and subsequent concentration of the standard upon repeated opening of the bottles containing the working standards. These solutions remained unchanged during a 6 month period at -10 to $15°C$.

Stock solutions are diluted by a factor of 200 to prepare working standards. These dilute solutions cannot be stored because of errors contributed by adsorption of chlorinated derivatives on the inside wall of containers. Detailed procedures for handling these standard solutions in gas chromatographic columns equipped with electron capture detectors are supplied. With the recommended techniques, the detection limit of pesticides in blood serum is 1 ng/ml.

Although routine measurements in clinical laboratories have become more reliable because of the availability of improved standards, many laboratory technologists are not fully aware of the factors that are a constant source of errors in their daily work.[31] A pamphlet[32] supplying complete detail for handling clinical laboratory standards and preparing stock and working standards solutions from these materials should be useful for reducing errors in many clinical laboratories.

In the preparation of stock standard solutions of glucose (10.0 mg/ml), 10.000 g of glucose* are weighed accurately and transferred quantitatively with 0.1% aqueous benzoic acid to a Class A (see NBS circular 602) 1000 ml volumetric flask.[33]† Gently swirl the flask to dissolve the sample and dilute to within 3–4 mm of the calibration mark. Place the volumetric flask in a constant temperature water bath at $20°C \pm 0.1°C$. After thermal

---

* NBS SRM41a or ULTREX® grade, J. T. Baker Chemical Co., Phillipsburg, N.J.
† See Chapter 5, Section I.A.4, for water quality.

equilibration ($\sim\frac{1}{2}$ hr) remove the flask and dilute to mark with 0.1% aqueous benzoic acid. Stopper the flask securely and mix by repeated inversion (at least 12 times). Transfer 120 ml portions of the stock glucose solution to thoroughly cleaned, dry amber glass bottles with screw caps provided with polyethylene liners. Label, date, and store in the freezer at −18°C. These stock solutions are stable for 3 years. To dilute these solutions, thaw the stock solution and place in a constant temperature water bath at 20°C ± 0.1°C until thermal equilibrium is attained.

V. STABILITY OF DILUTE SOLUTIONS

A. METHODS FOR STORAGE

Errors in trace analysis frequently result from changes in the concentration of standard solutions and dilute samples during storage. Because samples can not always be analyzed immediately after collection, the stability of the solution with time must be investigated to detect variations of concentration due to (1) evaporation or transpiration of the solvent, (2) adsorption of the trace element on the walls of the storage container, and (3) chemical changes such as precipitation or colloidal formation of the element under conditions of storage.

The results of a year-long study of evaporation-transpiration changes for many analytical reagents in different storage containers are reported in Table 7 and 7A.[34] During the investigation the containers were not opened. Aqueous solutions generally showed less than 1% change. However, 12 M HCl in glass containers and in all-plastic containers except Teflon (FEP) showed greater changes. Both aqueous and organic liquids stored in glass reagent bottles with conventional ground glass stoppers showed high losses. Volatile organic solvents were lost from all types of containers. Sealed glass ampoules were found to be most satisfactory for long-term storage, but high density polyethylene, polypropylene, or glass bottles fitted with tight sealing caps were satisfactory for periods up to one year. Evaporation was minimized by storing samples in a desiccator containing the same solvent.

Freezing samples of dilute solutions immediately after sampling retards adsorption and container contamination, but changes in the concentration of traces or major elements must be prevented. For example, seawater samples were collected in Teflon-lined Nansen bottles and immediately frozen in 500 ml polyethylene bottles. Upon thawing and comparing the solutions with standard solutions of ocean water, considerable reductions in salinity were observed.[35] The salinity in original seawater (34.655%) was reduced to 33.521% when filled polyethylene bottles were placed upright in a freezer. The salinity dropped to 26.271% when the bottles were placed upside down.

TABLE 7.   Overall Percentage Loss (or Gain) Over the 1 Year Period[a]

	Plastic Containers[b]					Container Type A[c]
Liquid	Type A	Type B	Type C	Type D	Liquid	
Water, 100 ml	0.35	0.26	0.32		Water, 100 ml	(0.16)
Water, 50 ml	0.37		(0.001)	0.07	12 M HNO$_3$	(1.55)
12 M HNO$_3$	0.03	0.01	(0.01)		1 M HNO$_3$	(0.29)
1 M HNO$_3$	0.23	0.47	1.05	0.09	95% EtOH	(0.01)
12 M HCl	1.34	1.02	1.53	0.24	Hexane	(3.24)
1 M HCl	0.23	0.18	0.44	0.07		
12 M HF	0.19	0.21	0.23	0.07		
1 M HF	0.40	0.39	0.32	0.26		
12 M NaOH	(0.05)	(0.15)	(0.09)	(0.02)		
1 M NaOH	0.23	0.33	0.96	0.12		
95% EtOH	1.80	0.81	3.27	0.02		
Ether	100.0					
Hexane	100.0	57.3	49.9	22.4		
Methyl-isobutyl ketone	13.2					
Kerosene	33.5					
Blank		0.00	0.00	0.00		

[a] Percentage change is calculated on the basis of 100 g of liquid. A value in parentheses indicates a weight gain.
[b] Stored in air between weighings, 4 oz. plastic bottles: type A = polyethylene, black screw caps with economical polyethylene liners; type B = polypropylene, waxed screw cap; type C = high density polypropylene, waxed screw cap; type D = Teflon (FEP), waxed screw cap.
[c] Stored in desiccator above similar solvent between weighings (Ref. 34).

The results were easily interpreted when it was observed that part of the contents had exuded past the cap in both cases. The explanation is based on the fractional freezing of seawater that occurs by freezing of pure water at the surface followed by progressive freezing of the remaining concentrated solution. When filled bottles are frozen in the upright position, expansion of the contents during ice formation can produce loss of brine by escape at the cap. In the inverted position, concentrated brine escapes. This problem is eliminated by using half-filled bottles, to ensure that no part of the sample is lost during freezing.

B.   ADSORPTION LOSSES

The stability of very dilute standard or sample solutions with respect to adsorption and desorption phenomena must be determined before long-term storage is attempted. Evaluation of literature citations of adsorption onto

TABLE 7A. Overall Percentage Loss (or Gain) Over the 1 Year Period

Liquid	Glass Containers[a]				
	Type 1	Type 2	Type 3	Type 1[b]	Type 3[b]
Water, 100 ml					0.20
Water, 50 ml	0.14	0.01	2.13		
12 M HNO$_3$	0.03	0.11	1.21		
1 M HNO$_3$					(0.05)
12 M HCl	1.71	1.72	3.93		
1 M HCl					
12 M HF					
1 M HF					
12 M NaOH					
1 M NaOH	0.17	0.01	0.60		
95% EtOH	1.00	0.09	3.72	0.08	0.48
Ether	17.9	100.0	100.0		
Hexane	3.43	2.50	11.5	0.44	1.22
Methyl-isobutyl ketone	0.70	0.51	2.72		
Kerosene	0.02	0.10	0.52		
Blank					

[a] 2 oz. glass bottles were used; storage in air between weighings; type 1 = inverted stopper, type 2 = greased inverted stopper, type 3 = standard glass stopper.
[b] Storage in desiccators above similar solvent between weighings.
Reprinted with permission, G. J. Curtis, J. E. Rein, and S. S. Yamamuro, Anal. Chem., **45**, 996 (1973).

the walls of surfaces is difficult. Ionic concentration, temperature, pH, and clean surfaces represent important parameters. Although standardized guidelines for predicting the behavior of ultratraces at the solution-container interface are unavailable, some general observations are useful to cite. Loss of traces by adsorption of ions on laboratory glassware has long been recognized as a problem in radiochemistry.[36-39] The adsorption of fission products[37, 39, 40] and chromate[41] on glass has been studied to establish the efficiency of cleaning procedures. A comprehensive treatment of the adsorption of ions from nitric acid solution and a discussion of associated analytical problems have been given by Starik.[42]

Highly sensitive radiotracer techniques have proved to be quite satisfactory for investigating chemical stability of solutions and adsorption losses. The isotopes ^{137}Cs, ^{90}Sr, ^{91}Y, ^{144}Ce, ^{140}Ba-^{140}La, ^{95}Zr, ^{131}I, and ^{106}Ru-^{109}Rh were employed to compare relative adsorption on borosilicate glass, polypropylene, and glass microscope slides.[43] Effects of pH, carrier concentration, and pretreating or coating procedures were investigated in detail. For most elements it was preferable to use borosilicate glassware rather than polypropylene. Robertson used radiotracer techniques to investigate the stability of seawater containing traces of Cr, Ag, Co, Cs, Cu, Fe, Hf, Sb, and Zn. Minimal loss of these traces at pH ≤ 1.5 was established.[44, 45]

The data in Table 8 can be used as general guidelines for judging the stability of solutions of Al, Ca, Co, Cr, Cu, Fe, Mg, Mn, Mo, Ni, Pb, Sr, Ti, V, and Zn stored in borosilicate flasks.[46] All the cations remained in solution at their initial level for at least 24 hr at pH 1.5. Elements present at their initial level after 24 hours remained stable for up to 4 weeks. Since in these studies all metals were present simultaneously, the results may not be valid for each metal separately. Similar studies on the stability of 11 additional cations (Table 9) also emphasize the necessity to acidify solutions to pH 2 or below immediately after collection.[47]

Several investigations of the stability of mercury solutions have been made. In 1969 Benes and Rajman[48] comprehensively studied the sorption and desorption on polyethylene of ^{203}Hg from aqueous solutions 10^{-8}–10^{-5} M in Hg(II). Adsorption from 10^{-8} M solutions is appreciable (30%) at pH 0 and up to 95% in the range pH 10–12.7. In more concentrated solutions the adsorption was less, but the age of the solution had a substantial effect on adsorption at pH 4.3–7.1. Adsorption of mercury on glass has also been studied.[49] The use of cyanide, iodide, or oxidants to prevent losses of

TABLE 8. Concentration of Ions in Solution (0.5 and 1.0 $\mu g/ml$ Measured at Various pH Values 24 Hours After Preparation of Solutions

Element	Initial Concentration ($\mu g/ml$)	Concentrations ($\mu g/ml$) at 8 pH values							
		1.5	3.5	5.0	6.5	8.0	9.5	11.0	12.0
Al	1.0	1.0	1.0	0.1	0.1	0.1	0.5	0.8	1.0
Mo	1.0	1.0	0.55	0.50	0.55	0.65	1.0	1.0	1.0
Pb	1.0	1.0	1.0	N.D.	N.D.	N.D.	N.D.	N.D.	N.D.
V	1.0	1.0	0.95	N.D.	N.D.	N.D.	N.D.	0.75	0.90
Ti	1.0	1.0	N.D.	N.D.	N.D.	N.D.	N.D.	N.D.	N.D.
Co	1.0	1.0	1.0	1.0	0.95	0.85	0.05	N.D.	N.D.
Cr	1.0	1.0	1.0	0.25	0.15	0.15	0.20	0.20	0.20
Cu	1.0	1.0	1.0	0.95	0.45	0.15	N.D.	N.D.	N.D.
Fe	1.0	1.0	1.0	N.D.	N.D.	N.D.	N.D.	N.D.	N.D.
Mn	1.0	1.0	1.0	1.0	0.90	0.75	0.05	N.D.	N.D.
Ni	1.0	1.0	1.0	1.0	0.90	0.75	0.05	N.D.	N.D.
Ca	0.50	0.50	0.50	0.50	0.50	0.50	0.45	0.35	0.35
Mg	0.50	0.50	0.50	0.50	0.50	0.50	0.45	0.05	0.02
Zn	0.50	0.50	0.50	0.50	0.45	0.25	N.D.	N.D.	0.25
Sr	0.50	0.50	0.50	0.50	0.50	0.50	0.45	0.40	0.35

Reprinted by permission, A. E. Smith, *Analyst*, **98**, 65 (1973). Copyright the Society for Analytical Chemists.

TABLE 9.   Concentration of Ions in Solution (0.2 and 1.0 μg/ml) Versus pH Measured 24 Hours After Preparation of Solutions

Metal Ion	Present Initially (μg/ml)	Concentrations (μg/ml) at 9 pH Values								
		1.0	1.5	2.0	3.0	4.0	6.5	8.0	9.0	11.0
Cd	0.2	0.2	0.2	0.2	0.2	0.2	0.2	0.06	0.04	0.06
Li	0.2	0.2	0.2	0.2	0.2	0.2	0.2	0.2	0.2	0.2
Au	1.0	1.0	1.0	1.0	0.95	0.95	0.7	0.3	0.3	0.3
Bi	1.0	1.0	1.0	1.0	0.45	0.45	0.45	0.3	0.2	0.25
In	1.0	1.0	1.0	1.0	1.0	0.6	0.25	0.2	0.2	0.25
Pd	1.0	1.0	1.0	1.0	1.0	0.9	0.85	0.7	0.7	0.7
Pt	1.0	1.0	1.0	1.0	1.0	1.0	1.0	1.0	1.0	1.0
Rh	1.0	1.0	1.0	1.0	1.0	0.85	0.80	0.60	0.2	0.1
Ru	1.0	1.0	1.0	1.0	0.3	0.1	0.1	0.1	0.1	0.1
Sb	1.0	1.0	1.0	1.0	0.75	0.75	0.7	1.0	1.0	1.0
Tl	1.0	1.0	1.0	1.0	1.0	1.0	1.0	0.75	0.6	0.5

Reprinted by permission, A. E. Smith, *Analyst,* **98**, 209 (1973). Copyright the Society for Analytical Chemists.

metallic mercury by evaporation from extremely dilute, neutral, standard solutions of mercury compounds has been recommended.[50, 51] Several workers report that 0.1 ppm mercury solutions in 1 N HCl, HNO$_3$, or H$_2$SO$_4$, as well as in a mixture of H$_2$SO$_4$ and KMnO$_4$, showed no appreciable change in ^{203}Hg activity within one week of preparation.[52, 53] On the other hand, very rapid loss (60%) of mercury from creek water samples occurred within 15 min of collection of the sample and storage without any preservative treatment in polyethylene containers.[54] Acetic acid-formaldehyde (10:1), a widely used preservative, was found to be ineffective for preventing loss of mercury. Rates of loss of this element from natural and distilled water samples stored in polyethylene, in polyvinyl chloride, and in soft glass containers have been reported.[55] Losses to all containers were appreciable but could be decreased by immediately acidifying with HNO$_3$ to a pH of 0.5[55]

The adsorption characteristics of traces of silver on various materials has been studied extensively. The adsorption of Ag from 1.0 and 0.05 mg/l solutions in the presence of 0.1 M Na$_2$S$_2$O$_3$, NH$_4$OH, ethylenediamine, and NaCl onto borosilicate glass was studied as a function of pH and time. Adsorption on glass, polyethylene, and silicone coated containers was less than 1% in the presence of thiosulfate or (ethylenedinitrilo)tetraacetic acid (EDTA). Similar studies have been made for adsorption onto Vycor, polypropylene, polystyrene, and Teflon.[56] Less than 1% adsorption on the containers over 30 days could only be ensured by complexing silver with sodium thiosulfate. An ion-selective electrode study of the adsorption of silver (2×10^{-6} M) from solutions placed in Pyrex, Desicoted Pyrex, polyethylene, Vycor, and Teflon containers has been reported.[57] Adsorption levels ranging from 28 to 48% at the end of 30 days increased in the order: Vycor < polyethylene $\approx$ Teflon < Desicoted Pyrex < Pyrex. In the absence of complexing ligands and in neutral solutions, losses of silver were less than 2% for periods up to 1 day for Teflon. However none of the containers could be used for long-term storage under these conditions.

Long-term storage of 50 μg/l solutions of Cr^{3+} in borosilicate glass and in polyethylene containers that were treated in five different ways before filling with the test solution has been investigated.[58] Tests over 30 weeks demonstrated that unused borosilicate glass vessels, when washed with demineralized distilled water alone or with 6 N HCl and 6 N HNO$_3$, followed by thorough rinsing with water, did not cause any depletion of Cr^{3+} in solutions over the entire period of storage. Similar results were observed for siliconized glass containers. However coating of containers in trace analysis is not recommended. Acid-washing (HCl and HNO$_3$) of polyethylene and polypropylene containers accelerated the loss of chromium, but acid treatment of borosilicate glass did not alter the stability of the solution. A reasonable interpretation is that acid washes etch polyethylene surfaces, thereby increasing adsorption sites on the expanded surface. When blood serum pools were stored at ambient temperature (25–28°C) or at 4°C in polycarbonate and polyethylene containers, the chromium exhibited less variation in polycarbonate containers.

Silver, lead, cadmium, zinc, and tungsten adsorption on borosilicate glass, polyethylene, and polypropylene container surfaces was recently investigated by Struempler.[59] Polyethylene containers did not adsorb cadmium or zinc. Acidification to pH 2 with HNO$_3$ prevented silver, lead, cadmium, and zinc adsorption on borosilicate glass surfaces. Silver was not lost from solution (0.5 ng/ml) at pH 2 in polyethylene containers, but protection from light was necessary. Surprisingly, there was complete loss of silver in 4 days, irrespective of acidity conditions, when test solutions were maintained in polypropylene containers. At pH 2, 10 ng/ml of lead could be maintained in solution beyond a 4 day period in borosilicate glass only. Solutions of cadmium (1 ng/ml) and zinc (100 ng/ml) showed no loss after 60 days of storage at pH 2 or 6 in polyethylene, but 15% adsorption of nickel occurred after a 30 day period on all container surfaces. Studies with ^{109}Cd aided the determination of the mechanism and the magnitude of cadmium loss onto soft glass and plastic containers:[60] Cd(OH)$^+$ rather than Cd^{2+} was absorbed onto glass, but no interaction was detected in the case of plastic containers.

The rate of loss of selenium from aqueous solution stored in various containers has been studied by use of radioisotopes.[61] In Pyrex beakers, selected to be representative of borosilicate glasses, 1 μg/g solutions at various pH values, (pH 1.4–7.0) show a maximum of about 1.1% loss after 24 hr. Nitric acid solutions were stable up to 15 days, since losses were only 1% or less. In a polyethylene bottle, however, losses of selenium were about 8% over a 15 day period, even for nitric acid solution.

Problems of stability of dilute solutions make it advisable to dispense accurately known small volumes of relatively concentrated stock solutions (≥ 1 μg/ml) or to dilute aliquots immediately before use. Thus the need to store dilute solutions for periods exceeding several hours is obviated.

VI. MEASURING SMALL VOLUMES OF SOLUTIONS

Skill in the use of techniques and devices for high accuracy measurement of submilliliter quantities of solutions must be acquired by the trace analyst. Often a suitable nanogram comparison standard can be obtained only by dispensing microliter volumes of a standard solution containing a microgram per milliliter or greater quantity of the desired element. Errors in measuring this reference solution significantly affect the accuracy of the final result. A variety of micro- and ultramicro flasks, pipets, and burets for quantitative work are discussed below.

A. VOLUMETRIC FLASKS

The Committee on Microchemical Apparatus has provided guidelines for manufacture of small volume glass flasks and microliter pipets.[62] Specifications have been given for 1–5 ml volumetric flasks. Flasks of similar design containing 1 ml and weighing 19 grams have been manufactured with bases small enough to permit weighing on a chemical balance. Difficulties in transferring solutions from these flasks have been handled by designing special pipets with long, narrow delivery stems. With these pipets, withdrawal of all but a few hundredths of a milliliter of solution from the flasks is possible.[63] Polyethylene volumetric flasks (10 ml) are also available commercially, but no appraisal of their reliability has been made.

B. PIPETS

Manual and automatic pipets for measuring microliter volumes are available. The self-adjusting washout pipets (Figure 2) are quite popular but have poor repeatability because they retain considerable amounts of liquid, varying with viscosity and expulsion speed. However pipets of the design shown in Figure 2c are available in 1–4 and 5–500 μl sizes and have been used widely.[56] Capillary pipets of this design must be cleaned, rinsed, and

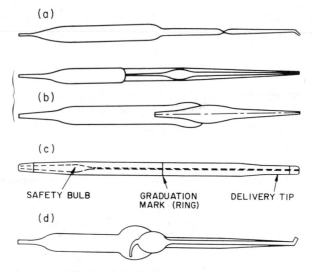

(a)

(b)

(c)

SAFETY BULB   GRADUATION   DELIVERY TIP
                     MARK (RING)

(d)

Fig. 2.   Self-adjusting washout pipets.

dried after each use and are not recommended for a busy laboratory.[64] Specifications have been given for Folin-type micropipets and for Pregl micro washout pipets.[65, 66] The 0.1 and 0.2 ml capacities of the former must be washed out to ensure complete delivery of the volume indicated by the graduation marks. Small amounts of wash liquid drawn up from the tip accomplishes this. To obtain quantitative delivery of the volume from Pregl-type pipets of 0.1, 0.2, 0.5, and 1 ml capacity, a small amount of wash liquid must be added from the top of the pipet.[67]

Because rinsing and drying of pipets is time-consuming, special pipets have been designed for use as serial dispensing or as convenient quantitative sampling devices with controlled rinsing (Figure 2d). This pipet permits more rapid and simple manipulation than has been accomplished with other types (Figure 2a).[68, 69] Self-rinsing of pipets, either with the sample itself or with a rinsing solution, allows fast pipetting of different samples without cross-contamination.[64] Transfer pipets can be rinsed conveniently by overflow, as demonstrated in Figure 3. The capillary (1) with both ends free is overflowed by suction and is rinsed by the overflowing portion of the sample. The main advantages of this design include volume dispensing with a high degree of repeatability ($\pm$0.1–0.3% standard deviation) for the range 0.05–250 μl, easy manipulation, and convenient calibration.[64]

The Sanz-type overflow pipet (Figure 4a) and a modified dispensing version (Figure 4b) are also convenient to use.[70] The former is activated by squeezing the polyethylene bottle and placing a finger over the vent hole.

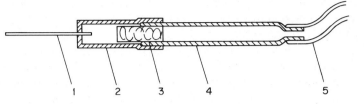

Fig. 3. Linderstrom-Lang Overflow Pipet: 1, plastic capillary; 2, transparent dome; 3, adsorption pad for overflow; 4, polyethylene tube; 5, flexible suction tube.

The polyethylene capillary then fills by suction, and the sample is dispensed by slowly squeezing the polyethylene bottle. The capillary is rinsed by overflowing liquid before the next sample is dispensed. For measuring solutions, the latter serves as an automatically filling polyethylene pipet with a reservoir. The solution to be measured is forced into the polyethylene pipet by squeezing the bottle and closing with a finger the vent hole in the upper, bell-shaped part. When the solution issues as a small drop from the outer end of the capillary, the opening is released and the pressure stopped. To release the sample, the vent hole is closed with the index finger, and the liquid is ejected by squeezing the polyethylene bottle.

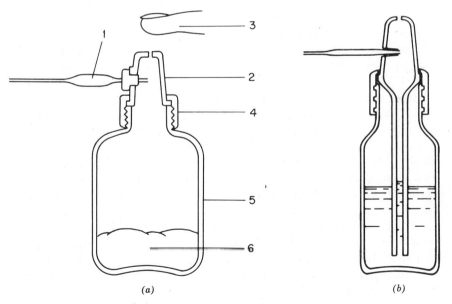

Fig. 4. (*a*) Sanz-type overlow pipet: 1, pipet; 2, holder; 3, finger; 4, cap; 5, polyethylene bottle; 6, Absorbent cotton. (*b*) Modified Sanz-type pipet.

Pipets that deliver a constant volume of liquid in a repetitive way from a reservoir—Linderstrom-Lang,[63] Grunbaum,[68] and Sanz types[63]—are especially suitable for quantitative work.

Piston pipets have the advantage that limiting the displacement of the piston provides semiautomatic measurements of different volumes with good reproducibility. Although recommended for volumes ranging from 1 to 200 μl, great loss of accuracy and repeatability occurs with volumes under 1 μl. Hamilton-type pipets,* with no dead space between piston and liquid, are now available with all-Teflon piston, barrel, and needles. Eppendorf† piston pipets with disposable plastic tips are available in 5, 10, 20, 50, 100, 200, and 1000 μl sizes. Displacement of the piston creates a vacuum that aspirates the liquid into the pipet tip. Calibrations of the 200 and 1000 μl sizes in this laboratory showed 0.3 and 0.2% relative standard deviations for the measurement of aqueous solutions. Deviations of ± 2 and ± 3 λ from the stated volumes occurred.

A multitude of commercial devices for measurement of microliter volumes have been introduced in recent years. Most are piston operated with disposable plastic tips. Since aspired volumes in piston-operated devices may vary with the density and viscosity of the liquid, with the angular position of the pipet, and with the immersion depth into the sample, these units must be carefully calibrated with the liquid to be measured. Disposable polystyrene pipets of 1 ml capacity graduated into 0.01 ml increments are also available. Their utility for high accuracy measurement of fractional milliliter volumes cannot be recommended.

Pipets for measuring volumes between 0.5 and 250 μl are available and volumes greater than 20 μl can be measured with an error of less than 0.2%.[71] The accuracy of volume measurements below this level can be improved significantly, however, by weighing the volume of liquid dispensed. Appropriate density-temperature relationships can then be used to compute the actual volume of solution dispensed. Microliter pycnometers (Figure 5) for this purpose are available in three sizes, 100, 40–80, and 10–30 μl.[67] The latter two sizes are graduated in 1 mm divisions and can be used even when the total amount of sample available is less than the maximum capacity of the pycnometer. Since the sample may occupy any portion of the graduated stem, the necessity of making a precise adjustment to a fixed calibration mark is eliminated. A modified Lunge-Key weight pipet (Figure 6) is reported to be useful when weighing a number of samples of a given liquid or when a specific sample size has to be used for a test.[72] The volume and number of graduations on the reservoir can be varied to fit individual requirements.

* Hamilton Co., Whittier, Calif.
† Brinkman Instruments, Westbury, N.Y.

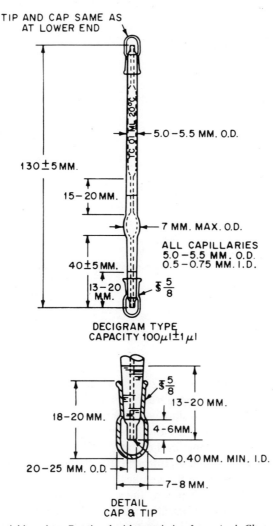

TIP AND CAP SAME AS
AT LOWER END

130±5MM.

5.0–5.5 MM. O.D.

15–20 MM.

7 MM. MAX. O.D.

ALL CAPILLARIES
5.0–5.5 MM. O.D.
0.5–0.75 MM. I.D.

40±5MM.

13–20 MM.

DECIGRAM TYPE
CAPACITY 100μl±1μl

18–20 MM.

13–20 MM.

4–6MM.

20–25 MM. O.D.

0.40 MM. MIN. I.D.

7–8 MM.

DETAIL
CAP & TIP

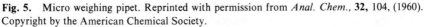

Fig. 5. Micro weighing pipet. Reprinted with permission from *Anal. Chem.*, **32**, 104, (1960). Copyright by the American Chemical Society.

Precision weighing pipets first used by Pregl[73, 74] for determining the specific weight of urine have been improved over the years and are regarded today as the simplest and most accurate devices for determining densities. Problems of vaporization and changes in the meniscus at different temperatures have been comprehensively studied and virtually eliminated.[75]

Appropriate microliter pipets often must be obtained by special fabrication. Prager and co-workers devised a system for the construction of pipets

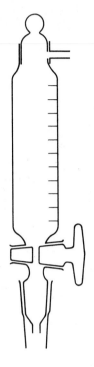

Fig. 6.   Lunge-Key weight pipet.

ranging from less than 1 nl to 200 nl volumes.[73] The finished product appears in Figure 7. This pipet was fashioned from a hypodermic needle whose hub was fitted with a silicone rubber septum cemented in place with epoxy resin. The needle formed the holder for a large bore tubing, which was inserted into a small hole in the center of the septum so that the tubing could be pushed through the septum and into the lumen of the hypodermic. The tubing fits snugly inside the number 17 needle; thus it remains parallel with the needle and is held firmly by the septum. Polyethylene tubing is slipped over the outside of the needle and is used to connect the needle and pipet to a syringe that provides the positive pressure needed to empty the pipet. By varying the length of quartz tubing (40–90 μm diameter), pipets of different volumes with reproducibilities of 1% over the entire range were made.

Polyethylene micropipets have also been conveniently constructed. Developed by Mattenheimer and Barner,[74] these pipets consist of a holder of glass tubing and the pipet proper of polyethylene tubing. To produce the pipets, a piece of polyethylene tubing is heated in the center over a flame until the material has softened sufficiently to be drawn so that the inner diameter of the polyethylene tubing is about 0.5 to 1 mm. The tension is maintained until the material has cooled.

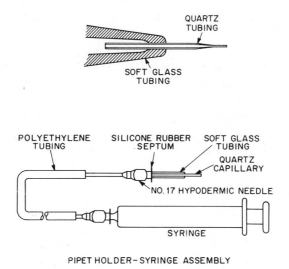

Fig. 7. Nanoliter quartz pipet.

The tip of the pipet (1–1.5 cm long × 0.1 mm diameter) is formed by cutting away the drawn portion with a razor blade. Depending on the inner diameter of the tubing and the desired volume, the necessary length is estimated and the tubing is cut.

A pipet holder is constructed by collapsing one end of a glass tube by slow rotation in a flame until the polyethylene pipet can just be inserted. Examples of finished pipets are presented in Figure 8. Polyethylene

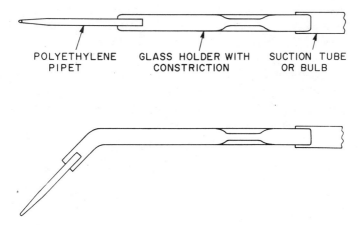

Fig. 8. Polyethylene micropipets. Reprinted with permission, H. Mattenheimer and K. Boiner, from *Mikrochim. Acta*, p. 918, 1959.

capillaries with inner diameter of 0.1–0.15 mm are used for pipets under 1 μl.

Prepared pipets are easily filled uniformly to the top rim by capillary action and can be blown out quantitatively. The hydrophobicity of polyethylene prevents fluid films from remaining and calibrations have shown excellent precision and accuracy:[74] Polyethylene pipets prepared in this way (0.4–10 μl) have been found to be as accurate as constriction pipets of much larger volume (50–100 μl). Thus for microchemical work, fabricated capillary pipets suitably calibrated are recommended over graduated capillary pipets or the commercially available constriction types.[74]

C. BURETS

Microburets with sensitivity of 10^{-9} ml have been constructed by using a plunger (0.008 cm diameter) in combination with a dial micrometer gauge having divisions equal to 0.00025 cm in linear traverse.* Calibration procedures for this device have been described.[75, 76] A convenient gravimetric microburet consisting of a small cylindrical vessel that rests on a balance pan has also been described.[77] The inlet and outlet connections of this device were constructed of thin-walled polyethylene capillaries that permitted 15–20 titrations per hour with an average consumption of 1 ml of titration solution. Special microburets have been fabricated for measuring oxygen-free or gas-saturated liquids.[78] Instruments of this type have proven to be useful for water determinations by the Karl Fischer procedure.

D. CALIBRATION

All micro measuring devices must be carefully calibrated before use in analytical work. Several methods can be adopted for calibrating a single device. Lowry[79] described a method in which aliquots of a highly colored solution are drawn by micropipets and the optical absorbance is determined after appropriate dilution. Comparison with standards similarly prepared with calibrated weight pipets provides the volume of the pipet.

In a similar procedure the volume is determined by pipetting 5.0 M $CaCl_2$ solution into 5.0 ml of 0.04 M HNO_3 and titrating with 0.1 N $Hg(NO_3)_2$ in the presence of sym-diphenylcarbazone as an indicator. Calibrations by fluorometric methods include dispensing with a nanoliter pipet an aliquot of a 0.1 N H_2SO_4 solution of quinine sulfate into 1 ml of distilled water.[75] The reading of this solution is compared with a standard curve obtained by diluting the stock solution to 1×10^6 or 10×10^6. By repeating the

---

* Emil Greiner Co., New York.

procedure several times for each pipet, the mean volume and standard deviation for each is computed.

Several methods in which mercury is dispensed by the micro device and weighed subsequently have been described.[80, 81] Pipets for measuring test samples (0.02–0.2 ml) have also been calibrated by transfer of a measured volume of mercury from a master pipet.[82]

Radioisotope techniques are rapid, can be automated, and are recommended when applicable. Suitable isotopes in appropriate solvents (similar to those used in actual analysis) are employed for calibrating to-contain or to-deliver pipets. In the case of piston-operated devices, however, aqueous solutions of suitable isotopes are preferred. Isotopically labeled solutions of a known number of counts per minute per milliliter are sampled several times with the calibration device. The count rate of the dispensed solution is then measured by liquid scintillation in the case of ^{32}P, a β-emitter, or with solid state detectors for γ-emitters such as ^{22}Na. Comparison of spectrophotometric and fluorometric calibration methods with radiotracer techniques have shown no significant differences.[83, 73]

Different pipetting methods have been compared by determining directly the volumes dispensed during calibration.[73] One method involved rinsing nanoliter pipets several times after delivery. In a second method the entire volume of the solution was emptied until an air bubble was formed; rinsing was eliminated. As expected, the apparent volume in the latter technique was slightly less, but the reproducibility matched the rinsing procedure. When the small amount of solution in the tip of the pipet was not rinsed or blown out, the apparent volume was less than that of the other two methods and the standard deviation was greater.

To obtain the highest reproducibility in delivering microliter volumes, several techniques have been recommended.[64]

1. The delivery should be viewed under microscopic observation to easily detect poorly dispensing devices.

2. The capillary tip should be wiped before and after aspiration of the sample by using prepurified cellulose paper moistened with solvent.*

3. To deliver into a liquid, introduce the tip of the capillary pipet, gently apply pressure until air bubbles appear, and withdraw while blowing. If a piston pipet is used, expel in air and remove the droplet from the tip by introducing the pipet into the liquid and executing a few waving movements, but do not stir.

4. To deliver on a surface with a capillary pipet, touch the surface in a nearly vertical position and apply pressure to eliminate air bubble forma-

* Purification of filter papers is discussed in Chapter 6.

tion. With a piston device, expel the liquid in air, touch the surface, and strip off the droplet adhering to the tip as described by Sanz.[64]

VII. SUMMARY

In recent years much progress has occurred in quantitative determination of ultratrace elements. This progress has basically resulted from the production and characterization of new trace reference standards and from the use of special techniques to reduce the size and variability of the blank. With appropriate standards, highly sensitive methods can be calibrated and the accuracy of trace determinations can be established. A shortage of appropriate reference standards and contamination problems will continue to persist. However synthetic standards often can be prepared, and the stability of these solutions can be checked. Control of contamination is not a hopeless task. Subsequent chapters of this book discuss the special care, patience, and facilities needed to make significant advances in preventing contamination from air, reagents, and containers.

REFERENCES

1. G. Tölg, paper presented at International Symposium on Microchemistry Technology, University Park, Pa., August 1973.

2. G. Tölg, *Ultramicro Elemental Analysis*, Wiley-Interscience, New York, 1970.

3. T. J. Murphy, paper presented at Seventh Materials Research Symposium, National Bureau of Standards, Gaithersburg, Md., October 7–11, 1974.

4. H. Kaiser and H. Specker, *Z. Anal. Chem.*, **149**, 46 (1956).

5. G. Tölg, *Talanta*, **19**, 1489 (1972).

6. J. P. Cali and W. P. Reed, paper presented at Seventh Materials Research Symposium, National Bureau of Standards, Gaithersburg, Md., October 7–11, 1974.

7. K. Heydorn, paper presented at Seventh Materials Research Symposium, National Bureau of Standards, Gaithersburg, Md., October 7–11, 1974.

8. W. H. Gries and E. Norval, *Anal. Chim. Acta*, **75**, 289 (1975).

9. W. H. Gries and W. L. Rautenbach, Sixth International Symposium on Micro-techniques, Graz, September 1970. Preprints, Vol. E., Verlag der wiener Medizinischen Akademie, Vienna.

10. J. W. Mitchell, E. Bloom, and C. Gooden, to be published.

11. J. W. Mitchell, C. L. Luke, and W. R. Northover, *Anal. Chem.*, **45**, 1503 (1973).

12. M. Zief, A. J. Barnard, Jr., and T. C. Rains, *Clin. Chem.*, **19**, 1303 (1973).

13. Report prepared by Analytical Standards Committee, *Analyst (London)*, **90**, 251 (1965).

14. W. G. Pfann, *Zone Melting*, Wiley, New York, 1966.

15. G. J. Sloan and N. H. McGowan, *Rev. Sci. Instrum.*, **34**, 60 (1963).

16. National Bureau of Standards, Notice on Trace Elements in Glass SRMs, SRMs 610 through 619, Office of Standard Reference Materials, Washington, D.C., July 1970.

17. T. E. Gills, W. F. Marlow, and B. A. Thompson, *Anal. Chem.*, **42**, 1831 (1970).

18. G. H. Morrison and N. M. Potter, *Anal. Chem.*, **44**, 839 (1972).

19. J. W. Mitchell, unpublished results.

20. National Bureau of Standards, SRM Notice on High Purity Platinum 680 and Doped Platinum 681, Office of Standard Reference Materials, Washington, D.C., December 1967.

21. National Bureau of Standards, Gold Standard, Office of Standard Reference Materials, Washington, D.C., September 1968.

22. National Bureau of Standards Reference Material 682, Office of Standard Reference Materials, Washington, D.C., June 1968.

23. National Bureau of Standards Reference Material 1577, Office of Standard Reference Materials, Washington, D.C., May 1972.

24. National Bureau of Standards Reference Material 1571 (Orchard Leaves), Office of Standard Reference Materials, Washington, D.C., February 1971.

25. *Chem. & Eng. News*, July 30, 1973, p. 12.

26. W. W. Meinke, *Mat. Res. Stand.*, **9** (10), 15 (1969).

27. Catalog of Standard Reference Materials, National Bureau of Standards Special Publication 260, July 1970.

28. 1973 Annual Book of ASTM Standards, American Society for Testing and Materials, 1916 Race St., Philadelphia.

29. *ASTM Standard. News*, **1** (7), 23 (1973).

30. J. F. Thompson, ed., *Manual of Analytical Methods*, Primate and Pesticides Effects Laboratory, Environmental Protection Agency, Perrine, Fla., November 1972.

31. F. W. Michelotti and G. C. Lindstrom, *J. Am. Med. Technol.*, **35**, 168 (1973).

32. G. C. Lindstrom, "How to Use ULTREX® Clinical Laboratory Standards in the Preparation of Stock and Working Solutions," J. T. Baker Chemical Co., Phillipsburg, N.J., 1972.

33. G. R. Cooper, *Workshop Manual of Methods for the Determination of Glucose*, Meeting of the American Society of Clinical Pathologists, Washington, D.C., September 18, 1966, p. 33.

34. G. J. Curtis, J. E. Rein, and S. S. Yamamuro, *Anal. Chem.*, **45**, 996 (1973).

35. D. E. Robertson, *J. Oceanogr. Soc. Jap.*, **28**, 37 (1972).

36. J. Belloni, M. Haissinsky, and H. N. Salama, *J. Phys. Chem.*, **63**, 881 (1959).

37. J. W. Hensley, A. O. Long, and J. E. Willard, *Ind. Eng. Chem.*, **41**, 1415 (1949).

38. A. O. Long and J. E. Willard, *Ind. Eng. Chem.*, **44**, 916 (1952).

39. G. K. Schweitzer and W. M. Jackson, *J. Am. Chem. Soc.*, **74**, 4178 (1952).

40. G. K. Schweitzer, B. R. Stein, and W. M. Jackson, *J. Am. Chem. Soc.*, **75**, 793 (1953).

41. E. P. Laug, *Ind. Eng. Chem.*, **6**, 111 (1934).

42. I. E. Starik, U.S. Atomic Energy Commission Report, AEC tr-6314, 1964.

43. G. G. Eicholz, A. E. Nagel, and R. B. Hughes, *Anal. Chem.*, **37**, 863 (1965).

44. D. E. Robertson, *Anal. Chem.*, **40**, 1067 (1968).

45. D. E. Robertson, *Anal. Chim. Acta,* **42**, 533 (1968).

46. A. E. Smith, *Analyst (London)*, **98**, 65 (1973).

47. A. E. Smith, *Analyst (London)*, **98**, 209 (1973).

48. P. Benes and I. Rajman, *Coll. Czech. Chem. Commun.*, **34** (1969).

49. P. Benes, *Coll. Czech. Chem. Commun.*, **35** (1970).

50. D. Lindstrom, *Anal. Chem.*, **31**, 461 (1959).

51. S. Shimomura, Y. Nishihara, and Y. Tanase, *Jap. Anal.*, **18**, 1072 (1969).

52. A. M. Igoshin and L. N. Bogusevich, *Gidrokhim. Mater.*, **47**, 150 (1968).

53. S. H. Omang, *Anal. Chim. Acta,* **53**, 415 (1971).

54. R. V. Coyne and J. A. Collins, *Anal. Chem.*, **44**, 1093 (1972).

55. R. M. Rosain and C. M. Wai, *Anal. Chim. Acta,* **65**, 279 (1973).

56. F. K. West, P. W. West, and F. A. Iddings, *Anal. Chem.*, **38**, 1567 (1966).

57. R. A. Durst and B. T. Duhart, *Anal. Chem.*, **42**, 1002 (1970).

58. V. D. Anand and D. M. Ducharme, AD-770021, National Technical Information Service, U.S. Department of Commerce, Springfield, Va., October 1973; V. D. Anand and D. M. Ducharme, Seventh Materials Research Symposium, National Bureau of Standards, Gaithersburg, Md., October 7–11, 1974.

59. A. W. Struempler, *Anal. Chem.*, **45**, 2251 (1973).

60. W. G. King, J. M. Rodriguez, and C. M. Wai, *Anal. Chem.* **46**, 771 (1974).

61. A. D. Shendrikar and P. W. West, *Anal. Chim. Acta,* **74**, 189 (1975).

62. Committee Microchemical Apparatus, *Anal. Chem.*, **28**, 1993 (1956).

63. Committee Microchemical Apparatus, *Anal. Chem.*, **30**, 1702 (1958).

64. M. C. Sanz, *Mem. Soc. Endocrinol.*, **16**, 27 (1967).

65. O. Folin, *J. Biol. Chem.*, **77**, 421 (1928).

66. J. Grant, *Quantitative Organic Microanalysis, Based on Methods of Fritz Pregl,* 5th ed., Blakiston, Philadelphia, 1951, p. 34.

67. Committee Microchemical Apparatus, *Anal. Chem.* **32**, 1045 (1960).

68. B. W. Grunbaum and P. L. Kirk, *Anal. Chem.*, **27**, 333 (1955).

69. P. L. Kirk, *Quantitative Ultramicroanalysis,* Wiley, New York, 1950.

70. M. C. Sanz, *Chimia (Aaran)*, **13**, 192 (1959).

71. A. Pasternak, Chem. Anal. (Warsaw) **13**(3) 593 (1968).

72. M. A. Birch, *Chem. Ind.* (*London*), 776, July 3, 1954.

73. D. J. Prager, R. L. Bowman, and G. G. Vurek, *Science,* **147,** 606 (1965).

74. H. Mattenheimer and K. Borner, *Mikrochim. Acta,* 916–921 (1959).

75. R. Gilmont, *Anal. Chem.,* **20,** 1109 (1948).

76. P. F. Scholander and H. J. Evans, *J. Biol. Chem.,* **169,** 551 (1947).

77. M. Stefl, *Chem. Listy,* **63** (10), 1142-8 (1969).

78. E. Scarano and M. Forina, *J. Chem. Educ.,* **47**(**6**), 482 (1970).

79. O. H. Lowry, N. R. Roberts, K. Y. Leiner, M. Wu, and A. L. Farr, *J. Biol. Chem.,* **207,** 1 (1954).

80. D. Francis, *J. Lab. Clin. Med.,* **22,** 718 (1937).

81. W. Lehmann, *Glas-Instrum. Tech.,* **1,** 8 (1957).

82. E. Newfield, *Med. Technol. Aust.,* **9,** 122 (1967).

83. J. F. Goggins and J. M. Tanzer, *Anal. Chim. Acta,* **45,** 526 (1969).

THE LABORATORY

Although control of temperature and humidity is routine in most analytical laboratories with sophisticated instrumentation, the necessity of regulating airborne contamination is not generally appreciated. Airborne contaminants in the form of dusts, mists, and fumes circulate in the atmosphere and enter laboratories through any vent.[1] The composition of the air in the laboratory, therefore, often approximates that of the surrounding atmosphere and fluctuates with the prevailing atmospheric conditions.

I. AIRBORNE CONTAMINATION

A. ANALYSIS OF PARTICLES

Aerosols composed of solid and liquid particulate matter are common constituents in ambient air. Other suspended solids are derived from natural sources (mineral dusts, pollen spores) and man-made contributions (manufacturing and distribution of industrial products, all forms of transportation). In the United States during 1957–1964, the concentrations of typical inorganic constituents of particulates in atmospheres were measured.[2] Samples collected during a 24 hr period showed a mean of 98 and a maximum of 1706 μg/m^3. The mean and maximum values in micrograms per cubic meter were as follows: sulfate, 9.3 and 95.3; nitrates, 1.6 and 24.8; iron, 1.9 and 74.00; lead, 0.5 and 17.00; zinc, 0.09 and 58.00. Up to 200 μg/m^3 of dust has been collected by filtering the air of a noncontrolled laboratory.[3] Analysis of this sample showed 10% Ca, 3% Fe, 1.5% Al, 0.5% Cu, 5% Li, 1.5% Ni, 1% K, 1% Mg, 0.5% Mn, and traces of other elements.

Since dust particles can weigh up to 1 μg and their composition changes with industrial and traffic conditions,[4] it is not surprising that the size of the blank in trace analysis can be correlated with atmospheric pollutants at different times. The activity of personnel in a factory raised the concentration of 0.5 μm particles from 0.2 $\times$ 10^6 at 10:00 A.M. to 1.5 $\times$ 10^6 during the height of activity at the noon hour.[5] In addition, smokers in a laboratory contribute to high blanks by circulation of smoke from one laboratory to another.

Small particles are also created by the operation of general air conditioning systems and machinery. Consequently the particle count indoors can be greater than the outdoor count.[6] Since suspended solids can easily contaminate samples, the blank is dependent on the history of operations previously performed in the laboratory. Chemical reactions, chemical spills, and corroded equipment can contribute to the formation of aerosols.

Rain or snow purifies the air to some extent; thus the size of the blank can be related to humidity.[7] Blanks taken in urban centers may be lower in the evening when traffic abates, population activities slow down, and air movement decreases. For these reasons the most sensitive measurements in some laboratories are taken in the evening by analysts when most laboratory workers are absent.[8]

B. CLASSIFICATION OF CLEAN ENVIRONMENTS BY PARTICLE COUNTING

Classes of clean environments are defined by the number and size of the particles present in the atmosphere. More than 50 different types of sampling instruments are available for the determination of particle mass and size.[9] Large particles (> 20 μm) are removed from laboratory air by sedimentation, whereas smaller ones (< 0.1 μm) tend to agglomerate. Usually airborne dust is a polydispersed sample with smaller size particles (1–3 μm) predominating. However a log plot of a normal distribution of particle size is a straight line.[10] A representative sample showed that 50% of the particles are smaller than 1.5 μm, but 50% of the mass distribution resides in particles larger than 7.8 μm.

Clean environments are based on particle control. In 1963 Federal Standard 209a was established to monitor the concentration of particles in the air of work areas.[11] Classes of clean environments were based on the maximum number of particles 0.5 and 5.0 μm in diameter per cubic foot of air. The definitions are listed in Table 1. Figure 1 plots the definitions relating particle size to airborne concentration.

In class 100, the cleanest environment specified, the greatest number of particles are below 0.5 μm in diameter, but the weight of contaminants in all the particles in this range represents less than 2% of the total weight of the particles. The definition of a class 100 environment in terms of 0.5 μm particles was imposed by the limitations in counting of air particles by light scattering techniques. The 0.3 μm range was the lowest range detectable when the definition was proposed.

C. REMOVAL OF PARTICLES

Attainment of class 100 conditions is based on air filtration. The most efficient filters are the dry type or viscous impingement fibrous filters. The

TABLE 1.   **Cleanliness Levels in Federal Standard 209a**

Class[a]	Maximum Contamination in Work Area (particles/ft³)
100	100–0.5 μm and larger
	0–5.0 μm and larger
10,000	10,000–0.5 μm and larger
	65–5.0 μm and larger
100,000	100,000–0.5 μm and larger
	700–5.0 μm and larger

[a] The standard requires laminar-flow equipment to attain this level of cleanliness. Since measurement of dust particles smaller than 0.5 μm introduces substantial errors, 0.5 μm has been adopted as the criterion of measurement.

dry type has proved to be the most feasible for the capture of small particles; it has been adopted for the high efficiency particulate air (HEPA) filter that is at the core of particulate control.

The HEPA filter was developed jointly by the Massachusetts Institute of Technology and Arthur D. Little & Company, Inc. for the Manhattan project[12, 13] These filters, designed specifically to prevent the discharge of radioactive dusts, are composed of thin porous sheets of ultrafine 100% glass fibers or a combination of glass and asbestos fibers. The sheets are pleated with aluminum or plastic separators between the folds to provide maximum area for filtering. The analyst must select a filter that contains plastic separators.* Several analysts who started with the conventional alu-

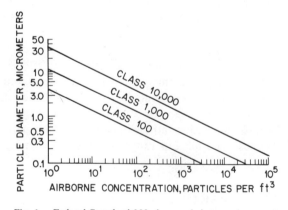

Fig. 1.   Federal Standard 209 classes of clean environment.

---

* HEPA filters with plastic supporting elements rather than aluminum are available from Air Control, Inc., Norristown, Pa.

minum separator have learned by experience that the acids used in wet ashing corrode the separators. Particulates from the filter then become a serious source of contamination. A standard filter, $61 \times 61 \times 14.9$ cm, assembled in a rigid frame of plywood, steel, or aluminum, provides a minimum airflow capacity of 500 ft^3/min. Standard HEPA filters have a minimum efficiency rating of 99.97% for 0.3 μm particles. The efficiency of the filters can be determined by weight, discoloration, or dioctyl phthalate smoke tests.[14]

Two types of air flow systems have been designed to remove particulate matter. The first is the conventional clean room that contains several HEPA filters spaced at intervals in the ceiling. The air is removed by return air grills located at the side walls at floor level. Because of the turbulent airflow conditions existing within such a facility, an airborne particle can pass a critical work area several times before leaving the work environment. These rooms usually meet the Federal Standard classes 10,000–100,000. Complete descriptions of class 100 conventional clean rooms built by industrial and governmental laboratories have been reported.[15]

The second type is the laminar-flow clean room derived from Whitfield's concept of "laminar" patterns of a unidirectional airflow.[16] Laminar-flow air moves in one pass from a bank of HEPA filters to the work area and exits from the clean room without any turbulence. When one wall of a room provides horizontal laminar flow, class 100 conditions are maintained in the work area close to the HEPA filters. When a HEPA filter bank is mounted in the ceiling, the entire room can be maintained at a class 100 level.

D. EFFECTS ON ANALYTICAL RESULTS

The need to control laboratory air during the characterization of ultrapure materials is documented by model experiments. When HF, HCl, and HNO$_3$ were analyzed for aluminum, iron, calcium, magnesium, lead, titanium, and boron in a laboratory atmosphere, the concentrations of these elements were an order of magnitude higher than in analyses carried out in a closed system containing inert gas.[7] Murphy[17] reported that the lead content in a conventional laboratory in Washington, D.C., was 0.77 μg/m^3. The first step in his preparation of low-lead mineral acids centered about the reduction of lead in the ambient air surrounding his distillation apparatus. Boutron[18] showed the effect of air purity in the analysis of a standard solution containing 10 ng/ml of iron. Values of 9.3 and 20.6 ng/ml were determined when solutions were concentrated in a laminar-flow hood and in the general laboratory atmosphere, respectively. Zief and Nesher[19] reported that the iron content of anhydrous sodium carbonate isolated from an ultrapure aqueous solution of the carbonate was 20 ng/g in a clean environment. When the solution was worked up in an open laboratory without regard to a clean air environment, the iron content increased by a factor of 2.

An unusual example of airborne contamination was demonstrated in the determination of lead in High Sierra lakes. One group of investigators reported the lead content to be 0.3 ng/g when the water sample was raised in a bucket to a helicopter. However the lead value dropped to 0.015 ng/g when a sample was meticulously taken aboard a small boat by a second team.[20] It is possible that the exhaust gases from the helicopter showered significant quantities of lead over the sample site.

II. CLEAN ANALYTICAL LABORATORIES

In the 1960s the high cost of class 100 laminar-flow clean rooms discouraged many applications. Today existing laboratories can be upgraded with minimum alterations and with no loss in existing bench and floor space.[21] To achieve an effective and economical conversion of a laboratory to class 100 conditions, a two-phase procedure is recommended.

A. UPGRADING THE CONVENTIONAL LABORATORY

1. Phase I

Phase I upgrades the general laboratory space. All incoming air from air-conditioning, heating, and ventilating sources must be provided with 85–95% efficient filters, a minimum approach to environmental control.[22] These filters are 85% efficient for 0.5–5.0 μm particles and 95% efficient for particles greater than 5.0 μm. The filters are placed behind an outer plastic grid inserted in a stainless steel frame, which is fastened to the air inlet with stainless steel screws. To minimize reduction in the linear air velocity from the inlet to the filter, the area of the attached filter housing should be considerably larger than that of the original air duct.

Upon installation of these filter boxes in our laboratory, unusually high accumulations of dust were detected in the air ducts. Evidently filters controlling the particles from outdoor air were not efficient. Consequently over a period of 20 years a high dust fall was observed in the ducts. After cleaning of the air ducts and installation of the filter boxes, the amount of dust on laboratory benches, shelves, and floors was notably lowered.

All walls should be coated with a resistant epoxy paint. Epoxy paints or styrofoam panels are recommended for the ceilings. Ready-to-use epoxy coatings are available along with two-component systems. Although the exact chemical composition of the catalysts and fillers is not generally printed on commercial labels, the two-component epoxy formula is preferred for clean rooms. Care should be taken to select a formula in which one component consists of a bisphenol A epoxy ether and a solvent cured by the addition of an amine or amide curing agent in a second solvent.* Pigment or tinting colors should not be added. Whenever any doubt exists about the elemental composition of paints or other materials, a sample should be submitted to the instrumental laboratory for analysis.

Although bench tops coated with epoxy paint provide a clean work surface, Teflon (FEP) or polyethylene sheeting is placed on the bench top for added protection. The Teflon is cleaned in aqua regia after each use; the polyethylene is discarded. Clean Pyroceram® (Corning) panels mounted on silicone supports also provide additional safety from bench contamination. Formica-covered furniture is also recommended, but this material stains easily when exposed to chemicals.

Floors should be covered with a vinyl flooring, preferably in one solid piece† rather than 23 × 23 cm panels, which accumulate dirt in the seams. To minimize dust accumulation, the vinyl covering should be coved at floor and wall intersections. To reduce traffic in and out of the laboratory, an intercom system and a pass-through port are provided for communication with workers and introduction of small samples from adjoining laboratories or corridors. All unnecessary shelving, partitions, and furniture should be removed. All windows and wall openings must be made airtight through caulking with silicones to prevent dust penetration. In addition, the room must be kept under constant positive air pressure of at least 8.4 m³/min. To eliminate chemical vapors in the general laboratory, an activated charcoal filter supplied with a blower system is placed in a strategic section of the laboratory.

2. Phase II

In Phase II sterile air is supplied to critical work areas via HEPA filters. Maximum usefulness of existing laboratory benches is attained by placing clean air modules approximately 1 m above the bench top. Dust control hoods and laminar-flow work stations for localized activities are recommended for the analytical laboratory.

An inexpensive vertical laminar-flow hood, 61 × 61 cm, has been found to be useful for sample preparation or transfer operations in the trace analytical laboratory (Figure 2). When turned on its side, it functions as a horizontal clean hood. This clean hood provides 0.36 m² of work surface.

---

* Tile-Clad II Enamel, Sherwin-Williams Co., Cleveland, Ohio.
† Vinyl Corlon, Armstrong Cork Co., Lancaster, Pa.

At the National Bureau of Standards one assembly of these hoods protects the integrity of ultrapure acids during purification by distillation.[23] Another assembly in the analytical laboratory (a three-unit clean air module) provides 0.72 m² for chemical operations normally performed in the open laboratory, as well as for a storage area for processed material.[24]

A medium to large (1.5–3 m) laminar-flow, full exhaust hood is essential for the trace analysis laboratory. Custom-designed units have been effectively used. By fitting one of the hoods shown in Figure 2 with a perforated Teflon base and placing it on a fiberglass plenum vented to a standard laboratory exhaust duct, 0.36 m² of clean fume hood was obtained. In another modification at J. T. Baker, a two-unit module received an internal coat of

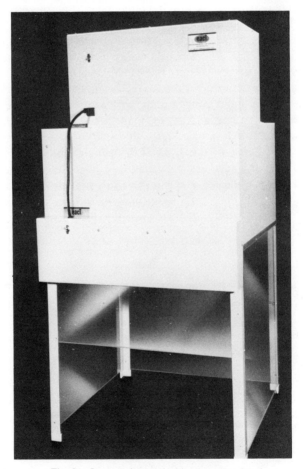

Fig. 2. Inexpensive vertical laminar-flow hood.

Fig. 3. Portable dust hood.

Kynar*. A 0.63 cm. perforated Teflon (TFE) base rested on a Teflon-coated plenum containing Teflon (FEP) tubing laced with 24 holes per foot and attached to a water source. The water spray in the plenum absorbs fumes liberated in wet ashing procedures. The water is collected in a trough connected to a standard laboratory drain. Adjustable exhaust louvers at the rear of the hood are vented to a standard exhaust fan. At Bell Laboratories an all-PVC, full exhaust, non-laminar-flow hood was designed for cleanroom work with highly corrosive chemicals that vaporize at ambient temperatures. The hood has an adjustable Plexiglas sliding front panel and all-PVC nozzles for regulation of water and gases. The hood can be placed on any flat-top laboratory bench or work desk. Laminar-flow exhaust hoods with perforated work surfaces are available commercially. These units have been used quite satisfactorily in the exhaust mode without detectable contamination of samples being evaporated.

Small, portable dust hoods (Figure 3) are the most economical sources of clean, non-laminar-flow air. The top housing contains a prefilter, a motor-blower assembly, and HEPA filter for 0.3 μm filtration. A clean hood fabri-

* Kynar is a polymer of difluoroethylene, made by the Pennwalt Corp., King of Prussia, Pa.

cated from 0.63 cm polymethyl methacrylate sheets requires no maintenance. One of these hoods, installed over a weighing balance in a spectrographic laboratory, has been in service for four years.[25] The value of the protective hood was shown dramatically on one occasion. While loading an elevator, a chemist broke a polyethylene container holding 11.3 kg of finely powdered alumina. The alumina was quickly circulated through the two-story research building by the air conditioning system. A fine powder settled on equipment, furniture, and floors. Several experiments in progress at the time were contaminated with aluminum, but the transfer of a sample to the crater of a carbon electrode under the dust hood was completed without contamination (aluminum was not detected later on the spectrographic plate).

A portable laminar-flow unit weighing only 13 lb* has been developed for moving analytical samples from a class 100 hood through a contaminated area to another class 100 location. In the pharmaceutical industry, drugs can be transferred with this unit through a penicillin dust atmosphere without cross-contaminating the sample.

For special analyses requiring limited work space, inexpensive, clean conditions can be attained with Plexiglas or polyethylene glove boxes that are continuously purged with pure air or nitrogen.[3, 7] Inflatable transparent polyethylene glove bags† are often advantageous, but the surface of the plastic itself can be a source of particulate matter.

When phases I and II are completed, the clean space under the HEPA filters is a class 100 area; the remainder of the laboratory becomes better than class 10,000.

B. POSITIVE PRESSURE NON–LAMINAR–FLOW LABORATORIES

Relatively inexpensive and efficient non-laminar-flow clean rooms can provide the analytical chemist with the necessary environmental conditions for performing trace measurements. Figure 4 is a schematic diagram of one of the clean facilities for trace analysis at Bell Laboratories. The floors, walls, ceiling, and windows are sealed; the room is then maintained under positive pressure with a continuous supply of filtered nonlaminar air. Air for the clean room is provided by a variable speed, blower-motor assembly. The air is passed through a prefilter, then through a HEPA filter, before entering the room. The blower is operated at high speed whenever the hood in the clean room is operated in the exhaust mode. At other times the low speed maintains sufficient airflow to keep the room under positive pressure.

---

* Portovoid, Air Control, Inc., Norristown, Pa.
† I²R, 108 Franklin Avenue, Cheltenham, Pa.

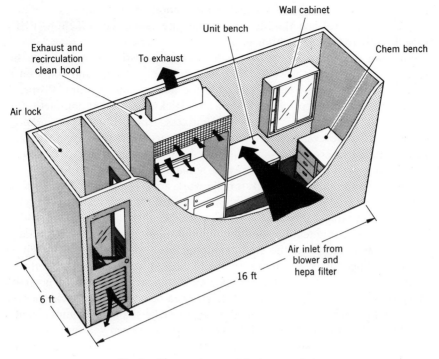

Exhaust and recirculation clean hood

Air lock

To exhaust

Unit bench

Wall cabinet

Chem bench

Air inlet from blower and hepa filter

16 ft

6 ft

Fig. 4. Clean environment for trace analysis.

By using the air lock to separate the clean facility from the remainder of the laboratory, the introduction of particulates from the noncontrolled area is minimized. The vertical laminar-flow clean hood, located inside the positive pressure room, further purifies the air of the laboratory and recirculates it over the work area inside the hood. The lower back portion of the clean hood contains an exhaust opening for venting noxious fumes. The fumes are vented to the outside air from a duct that carries the exhaust air from the bottom of the clean hood through an opening in the ceiling above the clean hood. When the exhaust lever is open, half the air entering the HEPA filter in the clean hood is recirculated into the general laboratory; the other half is exhausted to the outside. When the exhaust lever is closed, all the air enters the general laboratory. This type of hood is not recommended for open operations generating acid fumes, since these vapors may not be completely confined to the hood even if the exhaust louver is in the open position. Indeed, close examination of an analytical laboratory containing one of these hoods for analysis of acids revealed that metal faucets, hinges, and fixtures in the general laboratory were corroded by acid fumes from the hood. Wet ashing can be performed in these hoods if the sample is confined in a covered vessel and the liberated fumes are vented via a Teflon (FEP) tube directly to the exhaust louver.

Entering the air lock, one first encounters a plastic foam mat for the removal of dirt from the soles and heels. A supply of hats and coats on a hanger (see Garments) is available for the analyst and visitors. Properly attired, the analyst steps on a 61 × 61 cm "sticky mat"* that completes the removal of tenaciously bound particulates from shoes. The mat is constructed of several layers of adhesive materials. When dirt no longer adheres to the layer, the top sheet is peeled off, exposing a new active layer. Since shoes are a major source of dirt, some analysts prefer to keep a special pair of shoes in the air lock for the work area. Slippers or booties that slip over footwear are adopted by some chemists. Other investigators prefer to install mechanical shoe cleaners rather than change footwear.

The air lock, frequently called the change room, invariably is too small. When space is available, the air lock should be constructed large enough to furnish storage space for items routinely needed in the clean laboratory. Shelves should be provided for packages of plastic gloves, clean polyethylene bags, and foam wipers. In addition, mixed-bed ion-exchange cartridges, tools, and cleaning equipment such as mops, buckets, and a vacuum cleaner should be stored here. The availability of a sink equipped with hot and cold water is worthwhile for cleaning all materials before transfer to the clean laboratory.

Now the analyst is ready to enter the small laboratory designed for one or two analysts. Wherever possible, metal must be eliminated, particularly in the critical class 100 work area of the clean hood. The inner and outer walls, as well as the work deck, should be constructed of synthetic polymers. Although polypropylene is popular, its flammability poses a safety problem. Ring stands must possess ceramic bases; their vertical metal rods are best protected by slipping loose-fitting polypropylene tubing over them. Plastic clamps for supporting apparatus are available in most catalogs today. Plastic (Teflon TFE, PVC, or nylon) screws, bolts, and nuts† are readily available for details of construction. Stainless steel sinks corrode within months by acids unless protected by heavy-duty polyvinyl chloride linings. Galvanized sheet metal precoated with a polyvinyl chloride formulation‡ eliminates metal contamination from partitions and other components of storage cabinets.

Alloys such as brass should be replaced by plastics, if possible, or by stainless steel; door knobs, particularly, must be evaluated when ultratrace

---

* Laminaire Corp., Rahway, N.J.

† Product Components Corp., Mount Vernon, N.Y.

‡ FreAir, Camden, N.J.

copper, zinc, tin, lead, and iron are a concern. When plumbing lines cannot be replaced by plastic, they should be sealed in a plastic envelope behind the laboratory benches. The replacement of metal faucets with plastic is strongly recommended. Metal fixtures in clean hoods at the National Bureau of Standards are enclosed in polyethylene bags. When the polyethylene envelope fits loosely, the opening or closing of valves is not impeded by the polyethylene. The problem of metal outlets for water, compressed air, gas, and house vacuum arises when a conventional laboratory bench is converted to a clean area by installing a laminar-flow hood over the existing bench. Since metal pipes and valves for gas and compressed air lines are potential sources of contaminants, they should be removed from bench tops before clean air equipment is installed. A better plan is the capping of unused lines at the service area to the laboratory and removal of excess piping from the laboratory itself. Control valves for water and vacuum lines should be moved outside the clean area, preferably below the work surface of the hood; electrical outlets, in addition, should be located outside the class 100 area.

Cylinders for high purity gases tend to be dust collectors and particle emitters in every laboratory. The best solution is to store the necessary battery of cylinders in an accessible location outside the general laboratory and pass a Teflon (FEP) distribution line from the cylinder through the laboratory wall to a control valve in a corner of the laboratory. From the control valve 0.64 cm o.d. polypropylene tubing ard tee-valves can provide any number of outlets. Each tee-tube is connected to a reducing valve and a 0.2 μm membrane in-line filter at the point of use. Frequent changes of cylinders become simple operations inasmuch as no disturbance is created in the laboratory. Mechanical pumps should also be placed outside the laboratory to avoid oil vapor contamination. The removal of cylinders and pumps from the laboratory makes additional valuable space available. A utility room above or alongside the clean room is suggested for this equipment.

The unit bench contains a water purification unit. Distilled water is passed through two mixed-bed ion-exchange columns and a submicron filter, then is distilled in a vitreous silica still. The distillate is stored in a 10 gal polyethylene reservoir. Purified water is removed daily from the spigot into a polyethylene bucket, poured onto the floor, and removed with a squeegee-type mop constructed from heavy gauge steel with rust-resistant plating.

A variety of measuring devices are available for monitoring particulate matter. A relatively simple, inexpensive approach involves a disposable device combining the functions of a sampling filter holder for collecting airborne contamination and a microscope slide for filter examination at magnifications to 100×.* A 10 l/min flow-limiting orifice is installed in the outlet of the holder and connected to a vacuum source capable of pulling at least 50 cm Hg vacuum. At the end of the sampling period the vacuum is turned off the the filter holder is transferred to a microscope for examination and particle counting.[26]

The general laboratory area (Fig 4) is class 10,000; monitoring of the air at the chemical bench, the unit bench, and the entrance door shows 600, 700, and 800 particles per cubic foot. These levels can be maintained by daily maintenance of floors with the aid of a vacuum cleaner connected to a wall outlet from a separate house vacuum source.†

The air pattern in the laminar-flow hood or the general laboratory can be examined by placing liquid nitrogen in a polyethylene beaker and observing the "smoke" flow at any location in the laboratory. Commercial "smoke sticks" based on titanium tetrachloride contaminate the laboratory with titanium and HCl. Drawing the beaker of nitrogen along the face of a HEPA filter will immediately show the airflow during actual working conditions. Liquid nitrogen also demonstrates the extent to which bulky equipment in a clean hood causes turbulence and potential contamination from air in the general laboratory atmosphere. The liquid nitrogen should be placed frequently at the bottom of the laboratory entrance door to verify that the laboratory is indeed under a slight positive pressure.

Various airflows have been used in clean facilities. In the aerospace industry, a typical laminar-flow installation is usually regulated at an air speed of 30.5 (±3) m/min.[16] The airflow is adjustable from 10 to 45 m/min. This high velocity airstream contrasts sharply with the 4.6 m/min vertical airflow in medical applications for clean rooms.[27] In a turbulence study with titanium tetrachloride smoke tests in vertical-flow velocities ranging from 3.05 to 19.8 m/min, no smoke particles were detected 0.6 m upstream at velocities of 6.1–19.8 m/min. At 3.05 m/min, counts averaged $18–25 \times 10^6$ smoke particles per cubic meter of air.[28] Even though 4.6 m/min appears to be a marginally safe flow rate, this low velocity dramatically removed airborne bacteria from an operating room during surgical procedures. Speeds below 4 m/min lower the efficiency of the HEPA filter. Fewer particles impinge on the filter medium; more particles can find their way through the maze of the filter. In addition, air currents from the general laboratory area encounter less resistance in gaining entry to the hood. Speeds higher than 30.5 m/min tend to produce electrostatic charges on polyethylene and Teflon containers (skin effect). An air pressure of 30.5 m/min offers less resistance than one experiences while walking.

* Catalog MC/1, Millipore Corp., Bedford, Mass.
† Spencer Turbine Co., Hartford, Conn.

C. OTHER FACILITIES AND CONSIDERATIONS

Suspension of a clean air module over the sampling area has provided clean air in a small 4.8 × 8 m X-ray fluorescence laboratory.[29] A vertical, compact clean air module,* 1.26 × 1.07 × 0.45 m, housing a prefilter, motor-blower, and HEPA filter, is suspended from the ceiling over the balance. Continuous operation of the motors in this unit and in the filter of an adjacent laminar-flow hood efficiently clean this isolated room by recirculating clean air. The total recirculation of air gradually decreases the amount of contaminants entering the prefilters and, consequently, the downstream side of the HEPA filters.

Horizontal laminar-flow units obviously are impractical in the analytical laboratory if fuming is performed. In this case vertical laminar-flow units with exhaust capabilities (a "push-pull" airflow) provide the highest efficiency. The arrangement in Figure 4 is excellent when samples are positioned very close to the exhaust louvers. If the exhaust is too powerful, contaminants from the ambient atmosphere may enter the work area. Smoke patterns with liquid nitrogen define the exhaust capability needed to remove toxic fumes. A simple expedient is to fasten a thin plastic or glass fiber filament with cellophane tape to the front of the clean hood. If the filament is drawn into the hood, the airflow is not balanced. The exhaust louver must then be positioned so that the filament is forced toward the open laboratory.

Vertical-flow units require a partial closure for the top of the front. Without a top front panel, a sudden flow of laboratory air can penetrate as much as 7.6–10 cm into the clean hood, because air emerging vertically from the HEPA filter has a weak outward flow. The longer the front panel extends from top to bottom, the greater is the outward flow at the bottom of the unit. Small openings at the bottom, however, restrict accessibility to the hood.

Elimination of the front panel on vertical units has been accomplished by Nesher,[30] who attached an anodized aluminum curved sheet containing many small holes close to the HEPA filter. This "divergent membrane" provides a strong outward flow of air close to the filter, thus eliminating the induction effect (drawing room air into the laminar-flow hood) of a purely vertical airflow and permitting complete removal of the front panels.[17] The divergence of the membranes at the front of the laminar flow hoods gives the airflow patterns shown in Figures 5 and 6. Directional air thus provides an air curtain with unrestricted access for processing samples.[30] Experience has shown that no drafts penetrate the work area at the top or bottom of

---

* Envirco Corp., Albuquerque, N.M.

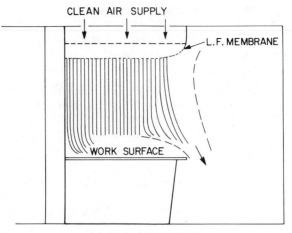

Fig. 5. Vertical laminar flow unit with diverging airflow for wall bench.

the hood. When spectrophotometers, balances, or other bulky equipment are employed, an anodized aluminum membrane curved 180° in the horizontal plane (Figure 7) produces converging laminar flow. These systems eliminate or substantially reduce turbulence around bulky obstructions.

Laminar-flow HEPA filtering units require proper maintenance. Imperfections in the filter medium and leakage through the gasket seal on

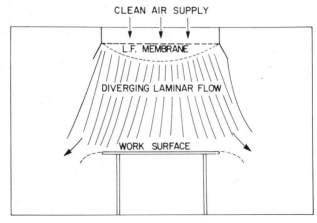

Fig. 6. Vertical diverging airflow for center bench.

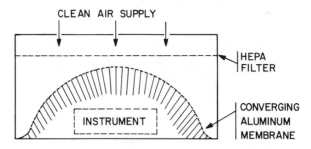

Fig. 7.   Converging airflow for bulky equipment.

the outer frame edge of the HEPA filter are the most important sources of leaks. Every joint in the filter housing is carefully sealed with silicones or neoprene by the manufacturer. As a result of careful quality control, one company has guaranteed a contamination level of less than 50 particles (0.5 μm and larger) per cubic foot, less than half the level specified in a class 100 environment.* Even though quality control has improved substantially in the past few years, the HEPA filter continues to be fragile. Improper handling during shipping can impair the effectiveness of a clean hood. A 100% scanning test of individual units, performed after installation, is the best assurance that the hood is in proper working order.

Prefilters should be changed after 3 or 4 months of continuous operation. The need to change the expensive HEPA filters is monitored by static pressure gauges that constantly record any drop in airflow rate. Overloading the filters with an accumulation of dust particles increases the static pressure. Four laminar-flow hoods in our laboratory have operated 24 hr a day for more than 20 months without a change of the HEPA filters. The prefilters were changed twice during this period. The static pressure increased from 0.75 to 0.92 in. of water. The HEPA filters and prefilters should be easily accessible, to prevent heavy dust spillage from occurring during replacement of these units.

It is good practice to check the performance of laminar-flow hoods at 6 month intervals. The HEPA filters should be scanned with an aerosol photometer to assure the analyst that no pinhole leaks exist. In addition, the particulate and air velocity levels should be monitored. A controlled environment particle analyzer† is a good investment when a number of clean stations must be checked.

* Westinghouse Electric Corp., Environmental Systems, Pittsburgh, Pa.
‡ Climat Instruments Co., Sunnyvale, Calif.

D. ELECTROSTATICS

Control of electrostatic charge is especially important in trace analysis, since plastic equipment is employed wherever possible to minimize trace element contamination. When materials such as glass, paper, cotton, steel, and various plastics are arranged in a triboelectric series, polyethylene and polyfluorocarbons are at the bottom of the list.[31] The fluorine atoms in polymers derived from tetrafluoroethylene, difluoroethylene, fluorinated ethylene-propylene, or trifluorchloroethylene confer an immediate negative charge to the surface, of the polyfluorocarbon when in contact with materials above it in the series.

The removal of a screw cap on a Teflon (FEP) bottle produces a negative charge on the neck of the bottle. Any particles attached to the polyvinyl chloride gloves or the nylon uniform of the analyst can be immediately transferred to the neck of the bottle during the handling of the cap. The development of electrostatic charge in handling Teflon equipment has been detected with an electrostatic meter* in our laboratory. Trained analysts can observe with the naked eye movement of particles from clothing or gloves to Teflon containers or apparatus.

Static eliminators consisting of a series of static bars mounted in a plastic frame behind an open grill of grounded rods have found acceptance in laminar-flow benches.† A high voltage (>5000 V) applied to needles spaced along the length of the bars creates a high potential gradient at the needle points, causing the surrounding air to ionize and carry away the static charge. Although this type of static eliminator is effective for food, drug, and plastics handling, the introduction of metal bars is not recommended in the analytical laboratory. Moreover, this powered device may cause a fire or explosion when volatile solvents are present.

Another static eliminator is based on ^{210}Po which is a pure α-emitter.* The isotope is absorbed by, and sealed in, ceramic beads ranging from 20 to 80 μm in diameter. The beads are coated onto a thin metal foil that is mounted in a specially designed housing. As the α particles streak through the air, they act as self-powered ionizers of air molecules, which then become a conducting medium.

Of the variety of configurations available, a static eliminator bar 30 cm long is most convenient for the laboratory. Even though the bar is metallic, the effect of potential metal contamination can be minimized by keeping the bar at the outside edge of the clean hood. The bar is remarkably effi-

---

* 3M Co., 3M Center, St. Paul, Minn.
† Simco Co., Lansdale, Pa.

cient. The question of safety was answered by Atomic Energy Commission approval for general distribution.

A device for negatively charging air was suggested in the early 1950s. Ionized air was created by photoelectrically ejecting electrons from thin layers of silver and gold foil attached to a copper screen surrounding an ultraviolet lamp.* When the device is placed behind a HEPA filter, the ultraviolet lamp can destroy bacteria and viruses that can accumulate on the filter and can assist in ionizing the air. The device could be considered when traces of copper, silver, and gold present no problems.

E. HUMIDITY

At low relative humidity ($< 40\%$), airborne particles develop electrostatic charges. Since charged particles are adsorbed more easily by synthetic fabrics than by cotton, an advantage offered by nylon laboratory coats is the ability to collect charged dust particles efficiently. In a class 10,000 area these particles can become a serious contamination problem when the relative humidity decreases below 50%. Maintenance of constant humidity has been found to be an important factor for the control of particulate matter in solids or solutions. When explosive vapors are present, a minimum value of 30% relative humidity should be maintained to eliminate electrostatic charges that can initiate explosions.

F. THE ANALYST

It is important to remember that the analyst is a primary source of contamination. The analyte in the class 100 area must be protected not only from airborne contamination but also from the analyst. Transfer of particles from the analyst in a class 10,000 area to the class 100 work area must be minimized by observing some of the rigid disciplines enforced in the burgeoning clean-room industry. The analyst must wear gloves, a hat, and a laboratory coat to prevent air travel of particles from hands, hair, and clothing.

The contribution from handling equipment with bare hands is best appreciated from the analysis of dried human skin: zinc, 6 μg/g; copper, 0.7 μg/g.[32] In the analysis of hair from adults the means of the zinc, copper, iron, and lead concentrations were found to be 108.5, 18.2, 22.3, and 12.2 μg/g, respectively.[33] Hands contribute skin flakings, cuticle, dirt, perspiration, and skin oils. Skin flaking of the hands and face become severe prob-

* *Construction of Ion Generators and Measurement of Negative Air Ions,* Westinghouse Lamp Div., Western Electric Corp., Bloomfield, N.J., 1951.

lems after a sunburn. Bismuth has been found as a contaminant from lip-stick after the pipetting of reagents. In addition, zinc is a common contaminant from lipstick and face powder. Since the use of lipstick, face powder, hand lotion, and nail polish is restricted in trace analysis, women will find these regulations rather harsh. Persons having acne, allergy, scalp, or skin problems should not carry out analyses. Chronic problems such as postnasal drip exclude potential technologists from the performance of ultratrace analysis.

Traces of iron, copper, gold, silver, platinum, and chromium are derived from wrist watches, gold and silver rings, and other articles of jewelry. Smoking, of course, is never permitted during trace analysis. Smoke contributes boron, iron, and potassium to the analyte.

The analyst must accept certain rigid rules and regulations and conform to them. Food, pencils, erasers, papers, and cloths are prohibited in the work area. Soles and heels should be wiped with a damp, clean urethane foam wiper* to remove grit and dust before entering the laboratory.

G. GARMENTS

Cotton and linen clothing are avoided because they shed considerable lint and have low abrasion resistance. Controlled linting is characteristic of the strong, continuous filament synthetic yarns and thread available today. Three of the most popular fabrics are made of nylon, Dacron® polyester or Tyvek® (duPont's high density, opaque, linear polyethylene fiber). A blend of 75% Dacron and 25% high tenacity viscose rayon reduces the generation of static electricity. A variety of coats and hats constructed from these fabrics is available from suppliers to the clean-room industry.†

Gloves must be impervious to skin oils and perspiration. No powder or tissue can be used in packaging. Polyvinyl chloride, polyethylene, and monofilament nylon tricot gloves have been reported to provide excellent protection against physical and chemical contamination. Experience has shown that polyvinyl chloride gloves packaged in class 100 polyethylene bags afford the best protection for the analyst.

Disposable, nontalc gloves should be worn for all routine operations from the cleaning and drying of laboratory ware to the weighing of samples and all subsequent chemical procedures. The handling of wash bottles is particu-larly important. When conventional polyethylene wash bottles containing high purity water were used to wash down precipitates in Teflon beakers, the iron content of the precipitate ($\sim$ 15 ng/g) was increased by the number

---

* Laminaire Corp., 1573 Irving St., Rahway, N.J.

† Clean Room Products, Inc., 55 Central Ave., Farmingdale, N.Y.; Angelica Uniform Co., Crescent St., Long Island City, N.Y.

of squeezes from the wash bottle.[27] The iron could have been leached from the polyethylene or adsorbed from the bare hands through the container walls. When concentrated hydrochloric acid bottled by automatic machines or by hand was examined, amino acids were detected only in bottled acid subjected to manual operations.

When animal tissues are sampled for ultratrace elemental analysis, one investigator wears three layers of polyethylene gloves. He then dips his gloved hands directly into cold concentrated nitric acid while a second investigator assists him to rinse off the acid with high purity water.[22] This procedure is not recommended as a general practice because of the obvious risk involved in handling concentrated acid.

H. LAUNDRY

A supply of disposable garments* should always be on hand. In addition, appropriate garments for daily use are suggested. New garments should be laundered in a clean room laundry because lint or contamination may be present as a result of manufacturing, packing, handling, or shipment. Local contracts for weekly laundry service provided in a class 100,000 clean-room laundry or better, according to Federal Standard 209a,[34] are also recommended. Laundered items must be returned in sealed, class 100 polyethylene bags that conform to clean-room standards.

Since all synthetic garments generate static electricity, checks for accumulation of static buildup should be made on occasion. Knee-length coats with conventional pockets may begin to lint after protracted use. Continuous monitoring is necessary to decide when these garments should be retired from service.

I. SAFETY

Trace analytical laboratories are frequently small rooms with only one exit. Since windows and walls are carefully sealed, the analyst must avoid storage of large quantities of inflammable solvents that could cause a flash fire and seal off the exit. He must also realize that the components of his clean hood and many containers are flammable organic polymers. Teflon and polyvinyl chloride are nonflammable, but polyethylene and polypropylene do burn. In one clean laboratory a serious fire was traced to the ignition of polypropylene panels. In any case, fire extinguishers should be enclosed in special protective units, situated in strategic locations, and inspected at regular intervals. The best way to meet increasingly rigid safety

* Fashion Seal Disposables, 64 New York Ave., Huntington, N.Y.

Fig. 8. Air-purging unit.

codes is to provide an emergency exit door, if possible, in the wall opposite to the entrance.

The Occupational Safety and Health Act[35] of 1970 tightened safety practices dramatically. Concern about the toxic effects of particular chemicals has led to the introduction of an air-purging unit (Figure 8) featuring a HEPA filter and blower, but the blower is connected to pull air from the room over the work area and trap particulates on the HEPA filter. The technologist is thus protected from particulates liberated in any operation.

REFERENCES

1. *Chem. Eng. News,* July 19, 1971, pp. 29–33.
2. P. W. Morrison, ed., *Contamination Control in Electronic Manufacturing,* Van Nostrand Reinhold, New York, 1973, p. 241.

3. J. Ružička and J. Starý, *Substoichiometry in Radiochemical Analysis*, Pergamon Press, New York, 1968, pp. 54–58.

4. G. Tölg, *Ultramicro Elemental Analysis*, Wiley-Interscience, New York, 1970, p. 13.

5. Ref. 2, p. 245.

6. Q. R. Thomson, "Indoor Versus Outdoor Air Pollution," *Contam. Control— Biomed. Environ.*, **11** (11 & 12), 22 (1972).

7. I. P. Alimarin, ed., *Analysis of High-Purity Materials*, Israel Program for Scientific Translations, Jerusalem, 1968, pp. 1–31.

8. T. C. Rains, U.S. National Bureau of Standards, personal communication.

9. S. A. Roach, in *The Industrial Environment—Its Evaluation and Control*, U.S. Department of Health, Education, and Welfare, National Institute for Occupational Safety and Health, 1973, p. 145.

10. D. A. Fraser, in *The Industrial Environment—Its Evaluation and Control*, U.S. Department of Health, Education, and Welfare, National Institute for Occupational Health and Safety, 1973, p. 155.

11. Clean Room and Work Station Requirements, Controlled Environment, Federal Standard 209a, General Services Administration Business Service Centers, August 10, 1966.

12. J. A. Paulhamus, in *Ultrapurity: Methods and Techniques*, M. Zief and R. Speights, eds., Dekker, New York, 1972.

13. H. Gilbert and J. H. Palmer, *High Efficiency Particulate Air Filter Units*, TID-7023, U.S. Atomic Energy Commission, Washington, D.C., August 1961.

14. *Contamination Control Handbook*, NASA-CR-61264, NASA—George C. Marshall Space Flight Center, Alabama, February 1969.

15. P. R. Austin, *Clean Rooms of the World*, Ann Arbor–Humphrey Science Publishing, Ann Arbor, Mich., 1967.

16. W. J. Whitfield, "A New Approach to Clean Room Design," Sandia Corp. Report SC-4673 (RR), Office of Technical Services, U.S. Department of Commerce, Washington, D.C., March 1962.

17. T. Murphy, Seventh Materials Research Symposium, National Bureau of Standards, Gaithersburg, Md., October 7–11, 1974.

18. C. Boutron, *Anal. Chim. Acta*, **61**, 140 (1972).

19. M. Zief and A. G. Nesher, *Environ. Sci. Technol.*, **8**, (7), 677 (1974).

20. C. C. Patterson and D. Settle, Ref. 17.

21. M. Zief and A. G. Nesher, *Clin. Chem.*, **18**, 446 (1972).

22. *Ind. Res.*, May 1975, p. 53.

23. National Bureau of Standards, *Tech. News Bull.*, **56**, (5), 104 (1972).

24. J. K. Taylor, ed., NBS Technical Note 545, Government Printing Office, Washington, D.C., December 1970, p. 53.

25. A. J. Barnard, Jr., and E. F. Joy, J. T. Baker Chemical Co., private communication, 1973.

26. 1972 Annual Book of ASTM Standards, Part 8, ASTM F25-68, American Society for Testing and materials, Philadelphia.

27. L. L. Coriell, W. S. Blakemore, and G. J. McGarrity, *J. Am. Med. Assoc.*, **203,** (12) 1038 (1968).

28. E. E. Choat and J. C. Little, Fourth Annual Meeting, American Association for Contamination Control, Miami Beach, Fla., May 25, 1965.

29. J. Kessler, Bell Telephone Laboratories, Murray Hill, N.J., private communication, 1974.

30. A. G. Nesher, U.S. Patent 3,426,512 February 11, 1969.

31. J. J. Keers and R. J. Kunz, *Nuclear Products,* 3M Co., St. Paul, Minn.

32. H. Feuerstein, *Z. Anal. Chem.*, **232,** 196 (1967).

33. J. P. Creason, T. A. Hinners, J. E. Bumgarner, and C. Pinkerton, *Clin. Chem.*, **21,** (4) 603 (1975).

34. 1972 Annual Book of ASTM Standards, Part 8, ASTM F51-68, American Society for Testing and Materials, Philadelphia, Pa.

35. Occupational Safety and Health Act of 1970, Public Law 91-596, 91st Congress, S2193, December 29, 1970.

MATERIALS FOR CONTAINERS
AND APPARATUS

I. STORAGE OF CHEMICALS

A. CONTAINERS

In storing high purity chemicals or reagent solutions, the optimum container materials and conditions must be selected to eliminate volatilization or leaching from the container walls. When a dilute solution is stored for subsequent analysis, these problems and also losses of traces by adsorption on the container surface are important. This section treats the first set of problems; the second topic was discussed in Chapter 2.

Reagent grade liquids are commonly stored in narrow-mouth bottles with loosely fitting dust caps (Figure 1a). On the other hand, high purity chemicals are contained in borosilicate vessels with inverted stoppers (male standard taper joint attached directly to container; matching female joint becomes the stopper). These inverted stoppers have been adopted by the electronics industry to preserve high purity during transfer operations. As ultrapurity techniques become standardized, it is safe to predict that inverted stoppers will be required for glass containers in the analytical laboratory.

Two designs of vessels with modified inverted stoppers were described by Hetherington.[1] In Figure 1b the inner cone and cap were ungreased but the outer ground glass surfaces were sealed with vacuum grease. A high purity polyfluorocarbon grease* has been the lubricant of choice in our laboratory.

Sealed ampoules (seal replacing the inner joint; the elimination of grease from the outer joint) that are modifications of Figure 1b have been described.[2] The vessel with the rounded bottom (Figure 1c) can be evacuated for storage of solutions *in vacuo*. Alternatively, inert gases such as nitrogen or argon can replace air when oxygen, carbon dioxide, and water vapor attack sensitive, high purity chemicals. The effect of light can be minimized by storing clear glass containers in the dark; leaching from container walls is retarded by storage between 0 and $-30°C$.

A study of different methods of packaging liquid reagents showed inverted ground glass stoppers and polyethylene-lined screw caps to be superior to conventional ground glass stoppers with respect to evaporation losses

---

* Halocarbon Products Corp., Hackensack, N.J.

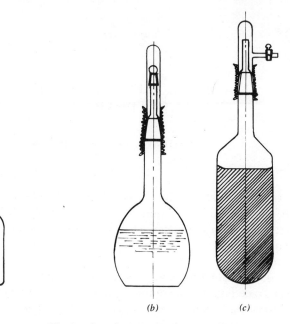

Fig. 1.   Containers for liquids.

from aqueous solutions and organic solvents (see Chapter 2). Greasing the inverted stoppers reduced evaporation losses of liquids of low volatility but enhanced the evaporation of volatile diethyl ether by improper reseating of the stopper after rising, due to internal pressure. Sealed borosilicate glass ampoules, of course, provide the greatest security against evaporation.

High density polyethylene, polypropylene, and Teflon (FEP) bottles fitted with tight sealing caps are satisfactory for storage of liquids. Screw caps with conical polyethylene liners significantly reduce evaporation losses. Leakage from caps on Teflon (FEP) bottles containing concentrated hydrochloric acid can be ascertained by placing the bottles in an outer polyethylene container on a laboratory shelf for a few weeks. When the outer container is opened, the odor of hydrochloric acid is usually strong.

B.   CONDITIONS

1.   The Laboratory

The conditions of storage and distribution are pertinent factors in the final purity of reagents. Unfortunately high purity reagents are frequently mishandled by the analyst. High purity chemicals should not be left in their original containers to collect dust on an open shelf or in the fume hood of a laboratory. The outside of the container should be kept properly cleaned to maintain the integrity of the contents.

High purity reagents and standard solutions may be stored at ambient temperatures inside a laminar-flow hood. Alternatively, primary containers of reagents and standard solutions can be placed in outer containers to provide excellent protection against dust contamination. Inexpensive household polyethylene juice dispensers found in most supermarkets have been used as outer containers for 1 liter Teflon (FEP) bottles.[3] Storage of the reagent in a class 10,000 area was then possible. Storage of liquids at low temperature ($-30-0°C$) in a refrigerator is preferred. Storage in the frozen state is a further refinement.

2. Commercial Production

In the commercial production of a reagent grade chemical, an aliquot is first submitted for analysis to a control laboratory. If this sample meets the predetermined specifications, the entire batch is packaged without subsequent analysis. In this case trace impurities contributed by the container are not determined. The trace impurity content of reagent chemicals is unimportant for many synthetic and analytical applications.

In contrast to the time sequence for the analysis of reagent chemicals, high purity liquids and solids are promptly packaged in suitable containers and are allowed to equilibrate with the container surface before the lot is finally analyzed.[4] Sensitive materials such as urea are packaged under argon or nitrogen; and they can be partially decomposed during the flame sealing of ampoules containing them.[5] Unless these chemicals are analyzed after flame sealing, their purity is not assured. The problem can be circumvented by packaging the labile material in a screw-capped vial which, in turn, is placed in an ampoule that is flame-sealed. Figure 2 shows the application of these techniques in the design of a container for purified bilirubin as a standard reference material for clinical use.

Teflon (FEP) containers are ideal for the transport of commercial high purity acids. However Department of Transportation (DOT) regulations do not permit the shipment of acids other than hydrofluoric acid in free-standing plastic containers. Vitreous silica and high silica (Vycor, Corning) glass have desirable chemical resistance, but economic considerations eliminate them from general adoption unless the chemical industry formulates a plan for return and reuse of the containers.

II. LEACHING

The literature on leaching trace metals from glass and plastics is difficult to interpret because fragmentary data on essential parameters are usually

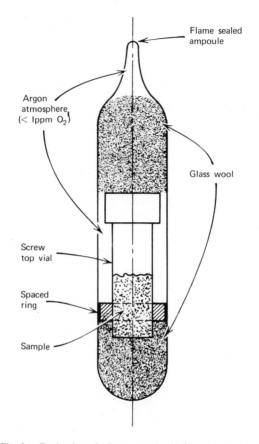

Fig. 2.   Packaging of a heat-sensitive high purity chemical.

reported. A complete description of the leaching action of a chemical on glass should record the composition of the glass, the upper temperature of glass working, and the manufacturer. The pretreatment of the surface with water, detergent, and hydrofluoric or other acid is also important. A description of the removal of surface contamination from polyethylene or polypropylene by abrasive action before acid cleaning may be the key to low desorption.

Access to radiotracer methods is frequently required to determine the origin and movement of traces. The leaching of arsenic from borosilicate glass was demonstrated as follows in the preparation of germanium tetrachloride free from arsenic. Good separation was achieved during fractional distillation in a borosilicate still, when the mole fraction of arsenic trichloride in germanium tetrachloride was greater than 10^{-5}. After distillation of original material with 3–5 μg/g of arsenic, neutron activation analyses of fractions detected very little reduction in arsenic.[6] When the arsenic was followed by radioactive [74]As, however, substantial fractionation was evident. Contamination by leaching of arsenic from the borosilicate glass vessel was eliminated by distillation in quartz. The distillate from the quartz apparatus was analyzed by neutron activation and radiotracer methods and was found to contain less than 6.3×10^{-10} mole fraction of arsenic.[7]

In many leaching studies on nanogram quantities of cations, contamination problems during handling or analysis invalidate the observed results. Aging or accelerated aging tests should be carried out to validate storage effects. For high purity chemicals normally stored in refrigerators, experimental storage at room temperature may be considered an accelerated aging test. In one laboratory accelerated aging tests at 50°C were carried out for high purity concentrated acids.[4, 8] The acids were sealed in 100 ml borosilicate ampoules designed to meet the drop test and other DOT regulations.[2] The acid was overlaid with a blanket of argon before sealing. The ampoule had a prescored breakoff tip to simplify opening. After opening, the standard taper joint of the dust cover fits tightly over the matched joint in the neck of the ampoule (Figure 3).

Preliminary evaluation of the ampoule was checked by filling with 5% HCl, sealing, and heating to 95°C. After an 8 hr exposure the ampoule was removed, drained, rinsed, and dried. A hazy ring on the inner surface at the sidewall just above the bottom indicated an area of poor durability.[9] The ampoule was then sectioned, and powder tests were made on each section. The glass was crushed to 40–50 mesh and allowed to react with 0.02 N H_2SO_4 at 90°C for 4 hr. Measurements of alkali extracted showed degraded durability near the bottom and the top of the container. With Pyrex 7740 degradation occurs during fabrication if the glass remains at a temperature between the annealing and softening points too long. Temperature adjustments during ampoule fabrication corrected this situation.

When the ampoules were properly constructed, aging studies on concentrated acids were undertaken. The leaching effects of dilute mineral acids,[10] but not concentrated acids, on borosilicate glass are recorded in the literature. A preliminary comparison showed that leaching is related to the water content of the acid.[4] Leaching decreased in the following order: 37% HCl > 70% HNO_3 > 98% H_2SO_4 > 100% HAc. Because hydrochloric acid exhibited the greatest leaching power, accelerated aging tests were conducted with this acid. Ampoules containing 100 ml of high purity hydrochloric acid were sealed and placed in a water bath at 50°C for up to 13 days. Actually, the bulk of the leaching was completed within 2 days at this temperature. Emission spectrographic analysis of the acid before and after aging at 50°C for 13 days showed that 12 elements (Sb, Ca, Cr, Co,

Fig. 3. Sealed ampoule for acids.

Fe, Pb, Mn, Hg, Ni, K, Sn, and Zn) remained unchanged.[8] Too frequently analysts dismiss borosilicate glass as an inferior container for acids. It appears that in trace analysis involving one of these 12 elements, high purity acids packaged in pretreated borosilicate glass can be used with safety. The only elements in the acid that were leached from the container were boron, silicon, aluminum, copper, magnesium, and sodium (Table 1). The increase in the sodium content far surpassed that of every other element. The

TABLE 1. Accelerated Aging Study of Concentrated Hydrochloric Acid in Glass Containers

Glass	Pretreatment	Storage Time (days at 50°C	B	Si	Al	Cu	Mg	Na
					Contaminants (ng/ml)			
Flint[a]	None		5	20	10	1	5	< 50
Flint[b]	Acid		5	30	10	3	5	80
Borosilicate[c]	Water	2.5	43	50	23	3	8.7	570
Borosilicate[c]	Water	13	67	57	27	27	8.0	630
Borosilicate[c]	Acid	2.5	8	30	10	3	7	90

[a] Freshly distilled acid at room temperature.
[b] 57 days at 0°C.
[c] Corning 7740 borosilicate glass.

original sodium count (< 50 ng/g) reached a level of 1200 ng/g when the ampoules were filled without any pretreatment whatever. Close cooperation with the manufacturer ensured avoidance of airborne contamination once the ampoules left the lehr at the glass-blowing facility.

Washing the ampoules with high purity water immediately before filling with acid reduced the sodium content to 630 ng/g after 13 days at 50°C. Leaching the ampoules for one week at room temperature with hydrochloric acid and thorough rinsing with water reduced the sodium to 90 ng/g, whereas the other five elements were almost unchanged. Special acid pretreatment makes possible the preparation of borosilicate A containers that do not add to the original impurity content significantly, except for sodium and silicon. Accelerated aging studies for 2.5 days at 50°C approximate levels reached at room temperature in approximately 6 months. Storage of commercial high purity acids in a cold room until shipment aids in preserving the integrity of the acid.

Neutron activation analysis detected readily leachable antimony in 1:1 HNO_3 after storage in vitreous silica at 80°C.[11] Under the same conditions in polyethylene, sodium was nonleachable, even though the Na content in polyethylene was found to be 208 ng/g. Manganese, copper, and chlorine, detected in polyethylene, also were not leachable.

III. SELECTION OF MATERIALS

An excellent review of materials suitable for containers and apparatus required for routine quantitative chemical analysis is available.[12] Detailed examination of the trace elements in some materials that are significant in trace analysis has also been recorded.[13] Additional considerations that can help preserve the integrity of samples during ultratrace analysis are presented in this section.

The problems contributed by walls of containers rival those induced by laboratory air in trace analysis. Walls can adsorb trace components from solutions, or traces can be leached from the walls by liquids. Both problems occur frequently. Surprisingly, ultrapure substances have been termed "universal solvents" not because of their high corrosiveness but because the reaction with container walls is ruinous to purity.[14]

Container materials have been rated in the following sequence for trace analysis applications: polyfluorocarbons > polyethylene > vitreous silica > platinum > borosilicate glass.[15] In general, this order of applicability reflects only the order of decreasing container contamination. The most important properties of these materials in the analytical laboratory are listed in Table 2.

Extensive experience in several laboratories attests to the advantages of Teflon, polyethylene, and vitreous silica. When Teflon (FEP) reagent bottles, beakers, and separatory funnels were offered commercially by laboratory suppliers, they quickly became standard equipment because investigators were anxious to minimize leaching from comparable items constructed from borosilicate glass. Teflon (TFE) provided excellent high pressure bombs for acid dissolution of samples. For acid treatment of materials at temperatures above 250°C, vitreous silica is the material of choice. Table 3 gives the most important uses for these preferred materials.

In addition to these materials, apparatus fabricated from other polymers or copolymers, ceramics, and a select group of metals is often indispensable to the analyst. As the level of trace constituents in a sample decreases, it becomes more and more important to evaluate the composition and properties of the container. Each group, therefore, is considered in greater detail.

TABLE 2. Properties of Container Materials

Materials	Temperature Limit (°C)	Chemical Resistance to			
		10% HF	10% HCl and 10% HNO$_3$	10% NaOH	Halogenated Hydrocarbons
Polyfluorocarbons	250	Excellent	Excellent	Excellent	Excellent
Polyethylene Linear	110	Excellent	Excellent	Good	Poor
Conventional	80	Excellent	Excellent	Good	Poor
Vitreous silica	1100	Poor	Excellent	Poor	Excellent
Platinum	1500	Excellent	Excellent	Excellent	Excellent
Borosilicate glass	800	Poor	Excellent	Poor	Excellent

TABLE 3. Preferred Materials and Their Use in Ultratrace Analysis

Teflon		Polyethylene (conventional)	Vitreous Silica
FEP	TFE		
Reagent bottles	High pressure bombs	Wash bottles	Evaporating dishes
Beakers		Storage containers for water	Pipets
Separatory funnels	Bottles for pressure filtration	"Clean" bags	Envelope for magnetic stirrers
Resin columns			
	Pressure filter apparatus	Bench top cover Filters	Scoops
			Dishes for microwave oven drying
	Beakers		Boats, muffles for furnaces

A. POLYMERS

1. Polyfluorocarbons

Polyfluorocarbons are stable up to 250°C, the upper temperature limit for organic polymers employed as container materials. Polyphenylene and polyimide polymers that can withstand temperatures of 500°C have been described but have not been fabricated into laboratory apparatus.[16] Polytetrafluoroethylene is the most chemically stable polymeric material available. Only molten alkali metals and elemental fluorine are reported to react with it. No organic solvents have been reported to cause distortion or swelling problems. Its disadvantages are permeability to gases and solvents, and poor thermal conductivity and cold flow. A disadvantage of Teflon (FEP) reagent bottles (Table 3) is the permeability to chlorine, hydrochloric, and nitric acids.

The unusually good chemical stability of Teflon (TFE) has been observed for Teflon (FEP) also.* Literature reports that both materials possess equivalent chemical resistance, however, are not entirely correct. Whereas 70% perchloric acid was unchanged after storage at −20°C for 6 months in 6 liter Teflon (TFE) bottles, the acid was contaminated with HF after storage under similar conditions in Teflon (FEP) blow-molded bottles.[17] When perchloric acid from each container was poured into borosilicate glass beakers, the beaker containing perchloric acid stored in a Teflon

* Teflon (TFE) and Teflon (FEP) are polytetrafluoroethylene and fluorinated ethylene-propylene, respectively, products of E. I. duPont deNemours & Co., Wilmington, Del.

(FEP) bottle was markedly etched within 10 min. The other beaker was unchanged. Teflon (FEP) contains a tertiary fluorine atom that is more reactive than the secondary fluorine in Teflon (TFE):

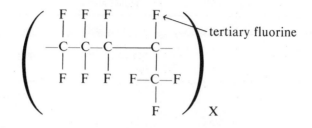

Teflon (FEP) bottles are extremely difficult to manufacture. Careful microscopic examination has shown the presence of many fine particles embedded in the walls about 7 μm below the surface. Laser probe spectroscopic analysis of these particles detected iron, zinc, aluminum, nitrogen, copper, and manganese among the constituents.[18] These impurities undoubtedly originate from the mold during fabrication of the bottle. They can be removed, in part, by prolonged immersion in boiling aqua regia. Until manufacturers can improve molding operations, these inclusions may play a role in ultratrace iron or zinc analysis. Thus far, the copious use of Teflon (FEP) bottles for ultratrace lead determinations has been fortunate, since this element is not a constituent of the metal inclusions. The bottles, in addition, have been provided with blue caps that contain 250 ppm of cobalt.[19] Strong acids and bases attack these caps and deposit an easily removable film on the lips of the bottles. For trace analysis the caps must be fabricated from virgin polymers.*

Conventional Teflon (TFE) tubing is somewhat stiff and awkward to handle. Flexible tubing consisting of an all-Teflon (TFE) film, fiberglass yarn, and an outer rubber latex sheath† affords a contaminant-free transfer medium for water, aqueous, or organic solutions.[20] A 90° bend radius for a 5 cm length of a 1 cm i.d. tubing demonstrates the unusual flexibility. The outer rubber sheath may be a source of trace zinc in the general laboratory environment. Loose-fitting polyethylene or corrugated Teflon (FEP) tubing slipped over the entire length of rubber can serve as a protective shield during the life of the Teflon tubing.

A 1:1 copolymer of tetrafluoroethylene and ethylene, Tefzel,‡ has better

* Discussion with the manufacturer led to the introduction of caps fabricated with virgin polymer.

† Gore-Tex, W. L. Gore and Associates, Newark, Del.

‡ Tefzel is a product of the E. I. duPont deNemours Co., Wilmington, Del.

mechanical properties than Teflon (FEP). Since Tefzel is easier to mold, bottles fabricated from it possess considerably less metal inclusions abraded from the mold. Since the 1:1 copolymer is approximately 80% TFE by weight, its resistance to 70% perchloric acid should be acceptable. Tefzel evaporating dishes usable with fluorides up to 180°C have been introduced.

Tefzel

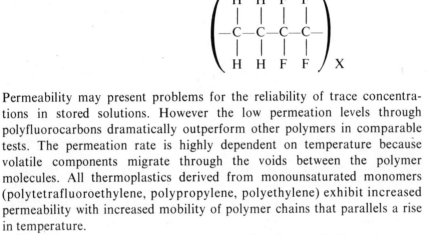

Permeability may present problems for the reliability of trace concentrations in stored solutions. However the low permeation levels through polyfluorocarbons dramatically outperform other polymers in comparable tests. The permeation rate is highly dependent on temperature because volatile components migrate through the voids between the polymer molecules. All thermoplastics derived from monounsaturated monomers (polytetrafluoroethylene, polypropylene, polyethylene) exhibit increased permeability with increased mobility of polymer chains that parallels a rise in temperature.

Oxygen or water vapor may enter from the air, or volatile components may diffuse from the inside of the container. Glacial acetic acid, hexane, benzene, carbon tetrachloride, and water permeate through Teflon (FEP) at room temperature and atmospheric pressure at a rate of 0.1–1.0 g/(24 hr)(100 in.^{2})(mil); oxygen and carbon dioxide permeate at 1.0–10.0 g/(24 hr)(100 in.^{2})(mil).[21] In Table 4 Teflon (FEP) shows the lowest permeability to carbon dioxide when compared with other materials.[22] Differences in wall thickness might cause variations on the order of 0.5–1.5 times the recorded values.

The microscopic voids that allow permeation also are responsible for absorption. This phenomenon is marked by a slight weight increase and occasionally by discoloration. For example, Teflon beakers containing urine samples are frequently discolored with various organic constituents, and the stains are not readily removable by extended cleaning.[23]

The surface free energy of Teflon is less than that of any other solid material. The nonadhesive surface is excellent for quantitative transfer. Removal of nanogram quantities of chromium and iron in suitably acidified solutions from a Teflon beaker via a hypodermic syringe and appropriate Teflon tubing is quantitative.[24]

TABLE 4.   Permeability of Polymers[a]

Polymer	Gases[b]			Water[b]
	Nitrogen	Oxygen	Carbon Dioxide	
Teflon (FEP)	21.5	59	17	500
Polypropylene	4.4	23	92	700
Polyethylene				
Conventional	20	59	280	2,100
Linear	3.3	11	43	120
Polyvinyl chloride	0.4–1.7	1.2–6	10.2–37	2,600–6,300
Polycarbonate	3	20	85	7,000
Silicone rubber		1000–6000	6000–30,000	106,000

[a] Units:

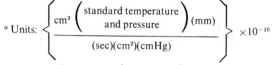

$$^a \text{ Units: } \left\{ \frac{cm^3 \left(\begin{array}{c} \text{standard temperature} \\ \text{and pressure} \end{array} \right) (mm)}{(sec)(cm^2)(cmHg)} \right\} \times 10^{-10}$$

[b] At 20–30°C.

Source: Ref. 22. Reprinted by permission of *Modern Plastics Magazine,* McGraw-Hill, Inc.

Polychlorotrifluoroethylene, Kel-F,* can be employed up to 180°C. The introduction of the chlorine reduces the resistance to chlorinated hydrocarbons, particularly chloroform and carbon tetrachloride. Polydifluoroethylene, Kynar, has a maximum working temperature of 150°C. Pinhole-free coatings of Kynar, applied to glass or metal, provide vessels with excellent chemical resistance.

A 1:1 alternating copolymer of ethylene and chlorotrifluoroethylene, Halar,† is a fluoropolymer easily processed by injection, blow molding, and extrusion. It is weldable and capable of continuous service at temperatures up to 180°C. The upper temperature limit can be extended to 200°C if the copolymer is cross-linked by high energy radiation. Sheets can be easily shaped into a variety of containers such as trays and boats.

2. Polyethylene

Conventional polyethylene is produced by a high pressure, noncatalyzed process at 150–250°C. The high density (linear) polymer is formed by a low pressure process catalyzed by oxides of transition metals or Ziegler catalysts. For applications in trace analysis, conventional, low density polyethylene is preferable to the high density form because of the former's

* Kel-F. is a product of 3-M Co., St. Paul, Minn.
† Halar is a product of Allied Chemical Corp., Morristown, N.J.

lower content of aluminum, chromium, cobalt, zinc, and titanium.[19] Polyethylene is the most common polymer in the analytical laboratory, mainly because of its low metal content, low price, and resistance to aqueous solutions of standards and reagents. Concentrated acids (other than hydrochloric) tend to react with it; aliphatic, aromatic, and chlorinated hydrocarbons produce softening and swelling.

Polyethylene is considerably more permeable to chemicals than Teflon. For example, diffusion of benzene and carbon tetrachloride is a hundredfold greater through polyethylene than through Teflon; the diffusion of water is tenfold greater. When a linear polyethylene bottle is filled with 28 ml of benzene and stored at room temperature, the benzene is completely gone after 4 or 5 months. In a conventional polyethylene bottle, the contents disappear within a month. Polyethylene obviously is a poor packaging material for a volatile organic such as benzene.

Low density polyethylene squeeze bottles for storing distilled or deionized water are common in most analytical laboratories. Chen[25] reported that fluorescent impurities were found in water stored for various times in polyethylene bottles. Old and new polyethylene bottles exhibited this effect. Although the impurities have not been identified, organic residues of polymerization initiators or fluorescent side products of the polymerization process may be involved.

When organic solvents such as low molecular weight alcohols were dispensed from low density polyethylene squeeze bottles in spectrophotometric and gas-liquid chromatographic (GLC) procedures, considerable irregularities were observed.[26] For example, reagent grade isopropanol, supplied commercially in glass bottles, exhibited surprisingly different absorbance curves from 260 to 330 nm when sampled from conventional squeeze bottles. The increased absorbance was ascribed to organic antioxidants, stabilizers, or antistatic agents added to commercial polyethylene. The level of antioxidants such as di-*tert*-butyl *p*-cresol in polyethylene is $60–600/\mu g/g$. The level of the antistatic agent Armostat 310* may be as high as 1500 $\mu g/$g. In addition to these additives in the polymer matrix, organic materials from the hands of the analyst or the laboratory bench permeate through the bottle. Upon concentration of a composite isopropanol sample from polyethylene bottles by a factor of 50, di-*tert*-butyl-*p*-cresol was detected on a GLC column (thermal conductivity detector) at a level of 25 $\mu g/g$ (or 0.5 $\mu g/g$ in the original isopropanol). Polyethylene containers must be scrupulously avoided in pesticide residue analysis for all samples examined by GLC instruments equipped with electron capture detectors.[27] Experience has shown that polyethylene and related plastics introduce

* Armostat, Armour Chemicals, Chicago, Ill.

serious interference when in contact with organic solvents, and materials such as sodium sulfate, glass wool, alumina, and filter paper employed in standard pesticide procedures.[28, 29]

Many aqueous solutions for analytical procedures are commercially available in polyethylene containers. In most cases appropriately cleaned polyethylene is not a serious source of contamination for inorganics. Examination of a solution of 1 M NaOH prepackaged in a polyethylene container, however, showed an intense absorption band at 410 nm. The polyethylene container appeared to be the source of the impurity because a 1 M NaOH solution prepared from reagent NaOH and stored briefly in a borosilicate glass volumetric flask showed no such impurity. Polyethylene should not be used for storing solutions that are capable of permeating the material. Other components may preferentially adsorb on the surfaces. For example, aqueous iodine solutions should never be stored in polyethylene containers because iodine is adsorbed.

In a test of the permeability of polyethylene, an iron rod was placed 5 cm away from a polyethylene bottle filled with 37% hydrochloric acid. Within a week the rod was attacked by hydrogen chloride that diffused through the container. Hydrogen chloride gas permeating through polyethylene obviously decreased the assay value of the acid. Solutions of ammonium hydroxide are similarly altered by diffusion. After 7 days at 63°C in a polyethylene bottle having a wall 1 mm thick, 30% solution of ammonium hydroxide lost 27% of the original ammonia.

3. Polypropylene

The chemical resistance of polypropylene approximates that of polyethylene. Polypropylene has slightly less resistance to sodium hydroxide, concentrated hydrochloric acid, acetone, amyl alcohol, butyl acetate, and phenol. Since it is harder and more rigid than conventional polyethylene, it is less permeable to volatile materials. An increasing assortment of labware is molded from polypropylene because its stability up to 135°C permits subjecting it to repeated autoclaving at 121°C. Polypropylene adsorbs trace quantities of most ions from dilute aqueous solutions more readily than glass.[30]

4. Silicones

Commercial silicones are, for the most part, siloxanes prepared by the addition of water to solutions of dimethyldichlorosilane in organic solvents. The resulting low polymeric silicones are converted to high molecular weight resins by the addition of metal salts such as zinc octoate.[31] Upon allowing 1 g of silicone rubber to be in contact with 100 ml of 0.1 M nitric acid overnight, 33 and 10 μg/g of iron and barium, respectively, were extracted. All other extractable elements were found to be below the microgram per gram level. The total elemental content found after ashing of silicone rubber was less than 100 μg/g.[32] Iron, barium, and zinc were extracted in greater amounts from natural silicone rubber than from polymethyl methacrylate, one of the purest polymers available. Frequently the extractable components of a material are not necessarily those present in greatest amounts. In polyvinyl chloride, for example, titanium is one of the principal contaminants, but dilute nitric acid extracts contain three- and twentyfold concentrations of iron and lead, respectively.[32]

Silicone rubber is one of the purest rubber materials for applications in the trace element laboratory. It shows small changes in properties over a wide temperature range (-70–$260°C$). Its excellent thermal resistance has been widely recognized in commercial coatings for laboratory heating tapes. Silicone rubber contains no plasticizer, sulfur, or other additives present in conventional rubber; thus it becomes the preferred material for stoppers in assembling special apparatus.

Silicone stopcock grease has displaced other lubricants for burets and separatory funnels. To remove silicone grease from the walls of glass equipment, the glass item is placed in concentrated sulfuric acid for a few hours, the acid is removed, and the glass is rinsed with water to remove the sulfonated silicone.

Foaming during distillations at atmospheric or reduced pressure can be eliminated by the addition of silicone antifoaming agents. Silicones are less volatile and purer than octyl alcohol, a popular defoamer. Silicones are effective at a concentration of 1 part in 100,000, thus providing a minimum of contamination. A common application for antifoam agents is distillation in the Kjeldahl determination.

5. Polymethyl Methacrylate

Since polymethyl methacrylate is derived from high purity methyl methacrylate by organic peroxide catalysis, it is relatively free from trace metals. Its transparency is an advantage in handling dilute acids and aqueous solutions of inorganic salts. Ion-exchange resin colums fabricated from this polymer preserve the purity of solutions undergoing purification. The polymer is satisfactory for alcohol-water mixtures at room temperatures but is not recommended for ketones and chlorinated hydrocarbons.

6. Polyvinyl Chloride

Polyvinyl chloride pipe and valves are standard equipment in many high purity water systems. These items have been reported to give many hours of satisfactory service in countless laboratories. PVC tubing, however, must be used with caution. Above 80°C the polymer is thermally unstable and liberates HCl. To combine with any free HCl and to increase the resistance of the polymer to heat and light, a variety of plasticizers and stabilizers are added.

More than 140 formulations of Tygon tubing are available, but only 8 are stocked as standard PVC tubings.* Solutions of samples in contact with this tubing may be contaminated with lead, titanium, zinc, tin, iron, magnesium, and other cations from the additives.

A PVC liquid formulation,† recommended originally as a safety coating for glassware, is suitable as a protective coating for racks, clamps, tongs, and other metal items in the laboratory. The object to be coated is preheated to 200°C in an oven, then dipped in the liquid to form a layer 1 mm thick, and cured at 200°C for 10 min.

B.   CERAMICS

Ceramics encompass all inorganic, nonmetallic products that are rendered serviceable through high temperature processing.[33] This definition broadens considerably the older view that ceramics pertain only to high temperature silicates. In the analytical laboratory silicates are still the major components of familiar pieces of apparatus, but oxides, carbides, and nitrides represent newer substances. In general, vitreous silica has displaced most other ceramic materials in the ultrapurity laboratory.

1.   Vitreous Silica

Vitreous silica, the purest form of silica available, is used frequently when freedom from contamination is essential. It is available as tubing, rods, and flasks, and can be fabricated into specialized equipment by competent glass blowers. Translucent (opaque) and transparent forms are made from natural quartz crystal powder or vapor phase hydrolyzates of pure silicon tetrachloride. If borosilicate is to be replaced by vitreous silica, the cost of the item increases by a factor of 10. All grades of vitreous silica are close to 100% purity as listed in Table 5.[34] Typical trace impurity levels appear in Table 6.[35] When silicon tetrachloride is the starting material, the residual chlorine content approximates 50 ppm.

Vitreous silica containers are good replacements for platinum for the ignition of most neutral and acid chemicals up to 1000°C. Alkali hydroxides and carbonates attack vitreous silica at elevated temperatures.

* Bulletin IT-130, Norton Co., Akron, Ohio.
† Plasti-Cote,® A. H. Thomas Co., Philadelphia, Pa.

TABLE 5. Major Chemical Constituents of Glass

Glass Type	Typical Concentration (wt %)[a]									
	SiO_2	Al_2O_3	ZrO_2	Na_2O	K_2O	Li_2O	B_2O_3	CaO	MgO	BaO
Soda lime A	73	1		17	0.5			5	4	
Soda lime B	74	2		13	0.5		3	11	0.5	
Borosilicate A	81	2		4	0.5		13			
Borosilicate B	73	6		7	0.5		10	1		2
Alkali-resistant	71	1	15	11	0.5	1				
Chemically strengthened	66	20		9[b]		5				
High silica	96	0.5					3			
Vitreous silica	100									

[a] In all except high silica and vitreous silica types F, Cl, SO_4, As, and Sb can be present in the range 0.05–0.5%.
[b] The surface is principally Na_2O.
Source: Ref. 34.

TABLE 6. Trace Impurities in Vitreous Silica

	Impurities (ppm)		
	Transparent		Translucent, from Fused Sand
Element	From Quartz	From SiCl$_4$	
Al	74	<0.025	500
B	4	0.1	9
Ca	16	<0.1	200
Cr	0.1	0.03	—
Cu	1	<1	—
Fe	7	<0.2	77
K	6	0.1	37
Li	7	—	3
Mg	4	—	150
Na	9	<0.1	60
P	0.01	<0.001	—
Ti	3	—	120

Source: Ref. 35.

High silica glass (Table 5) is also used for laboratory apparatus as well as ampoules for the distribution of acids.*

2. Borosilicate Glass

Glass is a noncrystalline material that is rigid from room temperature up to 800°C (borosilicate) or 1700°C (vitreous silica). Borosilicate A glass (Table 5) is commonly used for chemical labware such as flasks, beakers, and reagent bottles. In the United States, Corning 7740, Owens-Illinois KG-33, and Wheaton Vitro-200 are the three familiar code names for this formulation.

The low thermal expansion of borosilicate A glass permits rapid heating and cooling during analytical procedures, but sudden temperature changes must be avoided. When flasks heated to 200°C are transferred directly to a laboratory bench, drops of water on the bench will crack the flasks. This breakage can be eliminated by preparing a 7.5 × 7.5 × 1.25 cm solid block of Teflon (TFE), a poor heat conductor, for the transfer of hot dishes in a clean hood.

Borosilicate A can be used up to 500°C continuously or up to 600°C intermittently with solutions that are acid or neutral. A pH of 1 is only slightly more corrosive than pH 7. When water and ethanol were removed from silica gel by flash evaporation for 5 hr at 200°C in a 4 liter borosilicate A flask, the iron content in the silica increased only by 1–2 ng. The extraction of silica, sodium, and iron from glass containers is shown in Table 7.

Silicon represents a major component; sodium, a leachable component, and iron, a trace element. It is interesting to note that the leaching of iron in the presence of acid or water is the same for the three glass types. Attack by alkaline solutions is orders of magnitude greater than acid attack. Alkalies attack by etching the silica backbone and gradually solubilizing the glass according to

$$2X\ NaOH + (SiO_2)_x \rightleftharpoons XNa_2\ SiO_3 + XH_2O \qquad (1)$$

Acids leach the glass by exchanging hydrogen ions for the alkali ions present in the glass. Acid cleaning, therefore, not only removes contaminants attached to glass but also removes more active metallic ions from the glass surface itself, particularly at elevated temperatures.

Adams[34] has presented tables of corrosion studies that permit calculations of approximate contamination for chemical species as a function of

---

* Vycor (high silica glass) ampoules have been adopted by by G. F. Smith Co. for perchloric acid.

TABLE 7. Extraction of Elements from Glass Bottles[a]

| Glass | Element Extracted | Amount (μg element/ml solvent) | |
		Strong Alkali	Acid or Water
Borosilicate A	Si	1000	1
	Na	100	1
	Fe	1	0.001
High silica	Si	1000	0.1
	Na	1	0.001
	Fe	1	0.001
Vitreous silica	Si	1000	0.1
	Na	0.01	0.00001
	Fe	0.1	0.001

[a] 100 ml of solvent or solution in 100 ml flask for one year at 25°C.

temperature, time, and glass surface. Approximations for upper limits of contamination can be useful in determining the suitability of a particular glass container.

Borosilicate glass serves as a copious source of boron. A borosilicate bottle that held methanol for several months was shown by mass spectrometric analysis to possess 0.04 mole % trimethoxyborane $B(OCH_3)_3$.[36] Borosilicate is also a good source of boron, calcium, and cobalt for microorganisms. Vitamin B_{12}, for example, appears in water stored in borosilicate glass[37] but does not exist when the same water is held in polyethylene or polypropylene vessels. Borosilicate containers can vitiate studies to determine the nutrient value of trace elements. The discovery that vanadium[38] and tin[39, 40] are essential for the rat depended on plastic isolation chambers, filtered air, and purification of nutrients.

Unfortunately, some statements about the suitability of glass have been based on fragmentary data. One report states that borosilicate A glass is not suitable for trace analysis because a watch glass lost 1 mg SiO_2 in 60 min when water was boiled and 1.4 mg in 30 min when the water was replaced by ammonium hydroxide.[41] Since borosilicate A glass contains 81% silica, it is obviously not recommended for trace silica analysis. In addition, warm alkaline solutions are ideal for dissolving silica. Actually, accelerated tests conducted in this manner must be viewed with skepticism. Table 5 suggests that the leaching of trace elements such as iron or others that exist at the trace level in borosilicate A does not occur to any great extent at ambient temperatures. Even at 200°C the leaching of iron was found to be rather slow in our laboratory. When slurries of silicic acid were heated in borosilicate A flasks at 200°C for 8 hr, the iron content in 1 kg of product was found to be 10 ng/g. A 10 g sample was fumed with hydrofluoric acid; the residue was analyzed by emission spectrography.

A successful analysis may depend in large part on the previous history of glass equipment used in an analysis. New, unused apparatus is less reliable than old equipment that has been leached many times, particularly when the equipment has been used for one type of analysis only.

3.   Porcelain

Porcelain is a mixture of clay ($Al_2O_3 \cdot 2SiO_2 \cdot 2H_2O$), feldspar ($KAlSi_3O_8$), and quartz ($SiO_2$). Laboratory chemical porcelain is 48% SiO_2 and 49% Al_2O_3. Glazed and unglazed porcelain have working temperatures of 1100 and 1400°C, respectively. Glazes are low melting silicate glasses fused to the ceramic substrate; they consist of oxides of silicon, aluminum, copper, sodium, and potassium and are more stable to acids than to bases but unstable to hydrofluoric and phosphoric acids. It is interesting to note that the inside surfaces of mortars and the grinding ends of pestles are usually unglazed. Although porcelain crucibles, boats, and dishes enjoy widespread application in routine analytical operations, their usefulness is severely restricted in ultratrace analysis. Porcelain base support stands with aluminum alloy rods are excellent replacements for cast iron support bases with steel rods. The porcelain base, however, is usually mounted on four rubber feet. The rubber and the aluminum rod should be coated with a silicone coating compound,* yielding a film that resists corrosion, abrasion, and adhesion of foreign matter.

4.   Pyroceram

In 1957 the Pyroceram® family of glass ceramics was launched by Corning. This glass-ceramic invention, known to the consumer as CORNING-WARE®, has found applications in the laboratory. The new ceramic depends on a thermally induced process of heterogeneous nucleation and crystal growth. The final product contains more than 50% crystalline material after the addition of less than 10% of a nucleating agent.[42]

Pyroceram panels have found utility as wall panels in hoods and as bench tops. Rectangular panels up to 0.6 × 1.2 m provide an impervious work area comparable to glass in resistance to acids, alkalies, and organic solvents. The panels are supported by four silicone rubber feet; Pyroceram tops for magnetic stirrer–hot plates have excellent heat conductivity and

---

* Siliclad siliconizing fluid, A. H. Thomas Co., Philadelphia, Pa.

resistance to thermal and mechanical shock. In addition to these improved tops, the newer hot plates are supplied with noncontaminating nylon feet.

5. Graphite

Within the past 10 years pyrolytic graphite,* produced by chemical vapor deposition techniques, has become commercially available. It can be applied to conventional polycrystalline carbons or other substrates to reduce permeability and increase resistance to chemical attack. The object to be coated (e.g., a crystalline carbon crucible) or any porous substance that can withstand heating to 1050°C, can be placed in a high temperature vacuum furnace and contacted with toluene. The gaseous toluene "cracks," depositing pyrolytic carbon and forming hydrogen gas that is allowed to leave the furnace. The deposition of carbon is continued until the desired coating thickness has been attained.[43]

The resulting crucible can be operated at 2800°C under vacuum or in an inert gas atmosphere. Since toluene can be obtained very pure, the carbon deposited from this hydrocarbon becomes one of the purest ceramic materials available.

Pyrolytic graphite consequently is an ideal protective coating. It is resistant to attack by molten substances and has low thermal expansion. Its high electrical conductivity makes it a good static eliminator. Porcelain and Pyroceram cannot possibly reach the purity of pyrolytic graphite because these materials are based on powder technology. Powdered solids are extraordinarily difficult to isolate and store in the ultrapure state. Additionally, the conversion of the powder to a finished ceramic article by firing can only introduce further contamination.

6. Other Materials

A modification of porcelain is Mullite, an aluminum silicate offered commercially in the form of mortars and pestles with surfaces harder than porcelain. Equally hard surfaces are obtained in agate mortars fabricated essentially from quartz itself. Still harder surfaces are produced when 99.8% aluminum oxide is the sole component of mortars. Alumina withstands temperatures up to 1950°C and is inert to reducing and oxidizing atmospheres. Some resistance to alkaline and other fluxes is also provided.

Boron carbide (B_4C) and boron nitride (BN) are exceeded in hardness only by diamond. Boron carbide is synthesized by heating boron oxide, magnesium, and carbon. The product (m.p. 2400°C) can be shaped in gra-

* Pfizer Inc., 640 North Thirteenth St., Easton, Pa.

phite molds up to 2000°C, then machined with diamond. Boron nitride results from the reaction of boron trichloride and ammonia gases at elevated temperatures. As in the case of pyrolytic graphite, the purified gaseous reactants give an ultrapure product. Boron nitride and graphite are structurally similar. Layers of hexagonal rings of alternating boron and nitrogen atoms resemble the hexagonal rings of carbon in graphite. The similar structure parallels very high melting points, 3000 and 3300°C, respectively. When contamination control at high temperatures is a matter of concern, boron carbide, boron nitride, and pyrolytic graphite are the materials of choice.

C. METALS

Apparatus fabricated from platinum, aluminum, and stainless steel is observed in many analytical laboratories. When high temperatures are required for ashing, platinum crucibles are invariably used. Metal containers, however, are not recommended for ultratrace analysis. Wherever possible, vitreous silica or Teflon (TFE or FEP) should be employed.

1. Platinum

Platinum, unlike Teflon, is a good conductor of heat and can be heated safely to 1400°C. Since platinum conducts heat more rapidly than porcelain, it is the ideal material for rapid ignition of precipitates. Digestion with single mineral acids at elevated temperature can also be carried out conveniently. Heating any matrix with mixtures of nitric and hydrochloric acid in any proportion, however, can damage platinum. The ability to withstand HCl, HF, molten halides, and sulfates makes platinum an attractive container for these materials. The fused nitrates, nitrites, cyanides, oxides, and hydroxides of sodium and potassium, however, attack platinum.

Although platinum is considered to be inert in conventional analysis, small amounts of iron frequently contaminate sheet platinum. Differences in the trace metal contents of platinum from different sources proved to be the most important variable in the preparation of bulk soda lime glass from ultrapure raw materials.[44] In one series of experiments substantial differences of optical loss in bulk glasses depended on the source of platinum for the crucible in which the final melt was prepared. In another experiment an unusually high decibel per kilometer loss was traced to a copper contaminant in platinum.

Platinum for the laboratory is frequently alloyed with iridium or rhodium, since virgin platinum is too soft for most applications. Crucible alloy usually contains 0.3–1% iridium.[45] To avoid the formation of other alloys, platinum must not be heated in contact with any other metal.

2. Aluminum

Ultrapure gases are now shipped in aluminum cylinders with a special proprietary coating.[46] Determination of trace impurities in gases stored in these cylinders remains constant at the submicrogram level over a 12 month period, whereas the walls of clean, baked steel cylinders add impurities to a gas after it is pumped into the container. Wax cylinder-wall coatings for steel cylinders have not eliminated the pinhole problem.

Aluminum foil, washed with water and acetone, has been used for sample boats or liners for bottle caps. In view of the reaction of aluminum with common acid and base solutions, this practice is not recommended.

In the pH range 4.5–8.5 anodized aluminum (aluminum coated with alumina) affords excellent protection against corrosion. It is formed by passing a current through a cell containing an aluminum anode and a lead or graphite cathode immersed in an electrolyte. The thickness of the coating depends on the amount of electricity, the nature of the electrolyte, and the temperature.

3. Stainless Steel

Stainless steel Type 304 (18% Cr, 8% Ni) is widely used in the fabrication of laboratory electric ovens, baths, and furnaces.* It is important to remember that stainless steel is corrosion resistant, not corrosion proof. Corrosion resistance increases with the content of nickel; for example, Type 316 (16% Cr, 10% Ni) has higher resistance than Type 304.

The following materials attack stainless steel 304: hydrochloric, hydrofluoric, phosphoric, and sulfuric acids; halogen salts (chlorides, bromides, fluorides, and iodides); and sodium bisulfate. When several workers have access to a particular oven, therefore, it is important to restrict the chemicals or equipment placed in the drying chamber. Not only is there a danger of cross-contamination from the handling of different materials, but unexpected impurities in chemicals can be a source of corrosion. Stainless steel gravity convection ovens that have heating units exposed in the heating cavity should never be employed for drying chemicals or equipment. In our laboratory high chromium and nickel values after drying a calcium carbonate sample at elevated temperatures were traced to unshielded chromel resistance heaters in a furnace.

* Catalog SI-174, 1973, p. 198; Blue M Electric Co., Blue Island, Ill.

IV. CLEANING

In our laboratories technologists complain that an inordinate amount of time is devoted to the cleaning of laboratory ware. Certainly cleaning bottles for the collection of samples, plus preparing and storing reagents, are not glamorous aspects of research, but inattention to exhaustive cleaning of equipment vitiates considerable literature on ultratrace analysis. It is instructive therefore to describe in detail the sophisticated approach reported by Patterson and his group at the California Institute of Technology.[47, 48]

Items of Teflon (TFE and FEP) laboratory ware are totally immersed in aqua regia and cleaned by first heating at 55°C for one day. The color of the Teflon (FEP) should be yellow at this stage. The items are then totally immersed in high purity water for 1–2 days at 55°C, and in reagent grade concentrated nitric acid at 70°C for 3–7 additional days. At this stage the Teflon (FEP) becomes white (purple or yellow colors must be bleached out). The items are transferred to clean nitric acid in another tank for another 3–7 days of soaking at 70°C. At the end of this period rinsing with high purity water and soaking at 55°C for 3 days in very high purity water (specific resistance = 18 MΩ at 25°C; theoretical resistance for highest purity water = 18.3 MΩ at 25°C[49]) acidified to 0.1 wt % with ultrapure nitric acid. After soaking, the items are rinsed with water, placed in large Teflon (FEP) dishes, loosely covered with aluminum foil, and dried at 110°C. In a critical review of this procedure only the last step incorporating aluminum foil during drying can be considered questionable. Cleaning a 1 liter Teflon (FEP) bottle by this 21 day procedure has been estimated to cost $100—too expensive for adoption by commercial suppliers who offer marketable, ultrapure chemicals. Industrially, the same cleaning sequence is retained but in abbreviated form.

Comprehensive cleaning procedures must always be adopted for trace analysis. A container blank of 0.1 ng for a particular element is often needed. The cleaning operation, of course, must be designed to minimize the container blank with respect to the specific element of interest: lead, mercury, or arsenic (environmental laboratory); chromium, nickel, or zinc (clinical laboratory). Patterson, for example, described a procedure to obtain a container blank less than 0.1 ng of lead.[48] To an ultrapure solution in a "clean" container he adds a known amount of ^{208}Pb, heats for several days at 55°C, and analyzes the solution for extra lead by isotope dilution.

Cleaning steps vary with the chemical behavior of the element undergoing analysis. When lead is the element of concern, hot hydrochloric acid is an effective agent because it can form complexes with lead; nitric acid does not form complexes. Polyethylene collecting bottles for samples containing trace lead were cleaned by completely filling with analytical grade hydrochloric acid and allowing bottles to stand at room temperature for 3 days. The bottles were then rinsed with water and refilled with 1% ultrapure hydrochloric acid. After heating at 55°C for 3 more days, followed by rinsing with water and refilling with 0.1% ultrapure hydrochloric acid for 3 days at room temperature, the bottles were finally rinsed with water, covered by clean aluminum foil, and dried at 55°C in an oven.

Preliminary cleaning with aqua regia should be carried out in a conventional hood with a good exhaust system. The fumes should not be allowed to contact panels or equipment fabricated from stainless steel. The conventional hood should be placed alongside a polypropylene clean hood equipped with an exhaust system and a HEPA filter containing plastic dividers. Both hoods should possess sinks and water outlets to wash away acid spills. After the bulk of impurities is removed in the conventional hood, the final cleaning steps are transferred to the adjacent clean bench.

Cleaning procedures are performed in our laboratories as follows. Plastic containers (5 gal polyethylene drums with the tops removed) are first filled with an aqueous solution of a nonionic detergent.* Water with a specific resistance at 25°C of 18 MΩ/cm is conveniently obtained from tap water via a purification train of carbon, mixed-bed ion-exchange resins, and a 0.2 μm filter cartridge.† A satisfactory, alternative for high purity water is distillation in addition to the purification train just mentioned.‡ The detergent solution is best heated with an immersion resistance heater enclosed in a quartz sheath.§ Optimum effectiveness for cleaning is shown at 77°C. The analyst, wearing polyvinyl chloride gloves, then wipes the inner and outer walls of the apparatus briskly with polyurethane foam wipers, pours the detergent solution out, and rinses the vessel with high purity water to remove the last traces of the detergent. The empty container is then filled with a nitric acid-hydrochloric acid solution (16 ml concentrated HNO_3 and 48 ml concentrated HCl diluted to 100 ml with high purity water) and allowed to stand at room temperature. Polyfluorocarbons (Teflon TFE or FEP, Kynar, Kel-F) can tolerate a 24 hr contact with the acid solution without discoloration. Polyethylene becomes discolored after 8 hr; poly-

---

* Triton X-100, one of the octylphenoxyethanol series, a product of the Rohm and Haas Co., Independence Mall West, Philadelphia, Pa., has been found to have low metal content.

† Super-Q, Millipore Corp., Ashby Rd., Bedford, Mass.

‡ Barnstead, 225 Rivermoor St., Boston, Mass.

§ Thermal American Fused Quartz Co., Montville, N.J.

propylene develops color after 3 or 4 hr. Evidently the tertiary hydrogen in the polypropylene molecule is the point of attack for the acid.

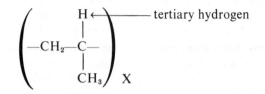

After the appropriate interval for leaching, the acid solution is discarded. The vessel must then be washed with copious amounts of water until the washings are neutral (pH meter or sensitive pH paper). Hydrochloric acid has been found to be valuable for iron removal because soluble complexes such as $HFeCl_4$ are formed. After the acid has been removed, the vessel is air dried in a laminar-flow hood or filled with the solution that is eventually to be added to the clean vessel. In general, after preliminary cleaning as described, the best leaching agent for a container is the solution that is to be stored in it.

A good general-purpose cleaning agent for glass as well as plastic is warm, concentrated nitric acid. A 5 or 6 hr contact time at 60°C, followed by copious washes with water having a resistance of 18 MΩ, and drying at room temperature, is satisfactory for glass and polyethylene. Polypropylene should not be heated with nitric acid at 60°C for longer than 1 hr. Ultrasonic baths are also excellent for borosilicate glass or vitreous silica but are ineffective for plastics.

Metals have been removed from glassware by washing in concentrated H_2SO_4-HNO_3 (1:1) and rinsing with high purity water. Here nitric acid has been substituted for the dichromate in the traditional mixture for cleaning glassware–sulfuric acid and an alkali dichromate. Since chromium is adsorbed[50] by borosilicate glass to the extent of 10 ng/cm² and the removal of the chromium by water washing has been found to be difficult,[51] dichromate is not recommended as a component of cleaning mixtures in the analytical laboratory. A mixture of 98% sulfuric acid and 48% hydrofluoric acid (1:1) is effective, but the contact time with borosilicate glass or vitreous silica should not be prolonged because hydrofluoric acid attacks these materials. Hydrofluoric acid and nitric acid mixtures also can be selected.

After the original washing in a detergent solution, the vessel can be immersed in 5–10% hydrofluoric acid for up to 10 min at room temperature, then rinsed carefully with water and air dried. Air drying, of course, assumes exposure of the vessel to air in a laminar-flow clean hood. Storage of a 5% solution of HF or 10% HNO_3 in a 5 gal polyethylene drum in a clean hood is a good replacement for the sulfuric acid-dichromate baths formerly used for cleaning in many exhaust hoods. The entire top of the drum should be removed so that large items can be handled.

Platinum crucibles should not be scratched with abrasive materials that may weaken the thin walls. Hot concentrated hydrochloric acid is the preferred cleaning agent for cation removal. Iron, a common surface contaminant, can be solubilized by complex formation. If hydrochloric acid is ineffective, digestion with nitric acid is a good second choice. If the acids are unsatisfactory, potassium pyrosulfate should be added to the crucible and fused at the lowest temperature possible for 5 min. Fusion with sodium carbonate has been suggested for removal of intractable materials, but attack of the platinum occurs when the crucibles are heated over a free flame or in a muffle.

Aluminum foil is soaked in redistilled acetone, rinsed several times with ultrapure water, dried at 150°C, and stored in a clean air environment.

Wrapping moist, clean articles in cleaned aluminum foil for oven drying is a popular technique, but ovens as well as the aluminum foil can be a source of contamination. An alternative is the construction of a pegboard from polypropylene panels and 15 cm lengths of polypropylene rods (1.25 cm diameter). The pegboard is supported from the top of a clean hood. Rinsed, clean laboratory ware is supported on the pegs for overnight drying (blowers on clean hoods in our laboratory have been running nonstop for 2 years). In addition, gloves, pipets, scoops, and so on, are stored on the pegs permanently, for cleanliness.

REFERENCES

1. E. F. G. Hetherington, *Zone Melting of Organic Compounds,* Wiley, New York, 1963.
2. R. E. McIlroy, U.S. Design Patent 216,362 (December 23, 1969), assigned to J. T. Baker Chemical Co., Phillipsburg, N.J. 08865.
3. K. Little and J. D. Brooks, *Anal. Chem.,* **46,** (9) 1343 (1974).
4. M. Zief and F. W. Michelotti, *Clin. Chem.,* **17,** (9) 833 (1971).
5. E. F. Joy, J. D. Bonn, and A. J. Barnard, Jr., *Thermochim. Acta,* **2,** 57 (1971).
6. H. J. Cluley and R. C. Chirnside, *J. Chem. Soc.,* 2275 (1952).
7. M. Green and J. A. Kafalas, *J. Chem. Soc.,* 1604 (1955).
8. N. A. Kershner, E. F. Joy, and A. J. Barnard, Jr., *Appl. Spectrosc.,* **25,** (5) 542 (1971).
9. P. B. Adams, Corning Glass Works, personal communication, 1971.
10. F. C. Raggon and F. R. Bacon, *Am. Ceram. Soc. Bull.,* **33,** 267 (1954).

11. J. W. Mitchell, J. E. Riley, and W. R. Northover, Technical Memorandum, Bell Telephone Laboratories, Murray Hill, N.J., 1972.

12. I. M. Kolthoff, E. B. Sandell, E. J. Meehan, and S. Bruckenstein, *Quantitative Chemical Analysis;* 4th ed., Macmillan, London, 1969, p. 452.

13. D. E. Robertson, *Anal. Chem.,* **40,** 1067 (1968).

14. P. Morrison, *Sci. Am.,* **228** (4), 122 (1973).

15. J. Minczewski, in *Trace Characterizations: Chemical and Physical,* W. W. Meinke and B. F. Scribner, eds., National Bureau of Standards Monograph 100, April 28, 1967.

16. N. W. Burningham and J. D. Seader, U. S. Clearinghouse. Federal Scientific Technical Information AD 1970, No. 711400.

17. M. Zief, J. T. Baker Chemical Co., unpublished results.

18. National Bureau of Standards Technical Note 459, D. J. Freeman, ed., December 1968.

19. D. E. Robertson, in *Ultrapurity: Methods and Techniques.* M. Zief and R. M. Speights, eds., Dekker, New York, 1972.

20. N. A. Karamian, *Am. Lab.,* **5** (12), 11 (1973).

21. E. C. Kuehner and D. H. Freeman, in *Purification of Inorganic and Organic Materials,* M. Zief, ed., Dekker, New York, 1969.

22. A. Lebovits, *Mod. Plast.,* **43** (7), 139 (1966).

23. P. O. Jackson, unpublished results, 1968.

24. J. Kessler, Bell Telephone Laboratories, personal communication, 1973.

25. R. F. Chen, *Anal. Lett.,* **5,** (10), 664 (1972).

26. B. M. Cumpelik, *Drug Cosmet. Ind.,* **113** (2), 44 (1973).

27. *Analysis of Pesticide Residues in Human and Environmental Samples,* J. F. Thompson, ed., U.S. Environmental Protection Agency, Research Triangle Park, N.C., December 1974, Section 2, p. 1.

28. D. F. Lee, J. Britton, B. Jeffcoat, and R. F. Mitchell, *Nature,* **211,** 521 (1966).

29. H. V. Morley and K. A. McCully, in *Methodicum Chimicum,* F. Korte, ed., Academic Press, New York, 1974.

30. G. G. Eichholz, A. E. Nagel, and R. B. Hughes, *Anal. Chem.,* **37** (7), 863 (1965).

31. R. R. McGregor, *Silicones and Their Uses,* McGraw-Hill, New York, 1954.

32. R. O. Scott and A. M. Ure, *Proc. Soc. Anal. Chem.,* **288** (1972).

33. W. W. Kriegel, in *Kirk-Othmer Encyclopedia of Chemical Technology,* A. Standen, ed., Vol. 4, 2nd ed., Interscience, New York, 1964, pp. 759–762.

34. P. B. Adams, in *Ultrapurity: Methods and Techniques,* M. Zief and R. M. Speights, eds., Dekker, New York, 1972.

35. G. Hetherington and L. W. Bell, in *Ultrapurity: Methods and Techniques,* M. Zief and R. M. Speights, eds., Dekker, New York, 1972.

36. R. P. Porter, *J. Phys. Chem.*, **61**, 1260 (1957).

37. S. H. Hutner, *Ann. Rev. Microbiol.*, **26**, 313 (1972).

38. K. Schwarz and D. B. Milne, *Science*, **174**, 426 (1971).

39. K. Schwarz, in *Newer Trace Elements in Nutrition*, W. Mertz and W. E. Cornatzer, eds., Dekker, New York, 1971, p. 313.

40. K. Schwarz, in *Trace Element Metabolism in Animals*, C. F. Mills, ed., Livingstone, Edinburgh, 1970, p. 25.

41. M. Knizek and J. Provaznik, *Chem. Listy*, **55**, 389 (1961).

42. U.S. Patent 2,920,971 (January 21, 1960), S. D. Stookey, Corning Glass Co., Corning, N.Y.

43. J. J. Svec, *Ceram. Ind.*, March 1971).

44. D. Pearson, Bell Telephone Laboratories, personal communication, 1973.

45. I. M. Wise and R. Vines, *The Platinum Metals and Their Alloys*, International Nickel Co., New York, 1941.

46. H. A. Grieco and W. M. Hans, *Ind. Res.*, **16** (3), 39 (1974).

47. Meeting Report, "Interlaboratory Lead Analyses of Standardized Samples of Seawater," *Mar. Chem.*, **2**, 69 (1974).

48. C. C. Patterson and D. M. Settle, Seventh Materials Research Symposium, National Bureau of Standards, Gaithersburg, Md., October 7–11, 1974.

49. V. Smith, in *Ultrapurity: Methods and Techniques*, M. Zief and R. M. Speights, eds., Dekker, New York, 1972.

50. E. B. Butler and W. H. Johnson, *Science*, **120**, 543 (1954).

51. E. P. Laug, *Ind. Eng. Chem.*, *Anal. Ed.*, **6**, 111 (1934).

CHAPTER 5

PURIFICATION OF REAGENTS

To solve the problems of modern trace analysis, rigid demands are imposed on the control of trace metals in water, acids, bases, solvents, buffers, supporting electrolytes, fluxes, oxidants and reductants, chelating agents, and in other chemicals required in analytical work. A target level of less than 10 ng/g is the objective for most trace impurities. Commercial suppliers have focused some attention on this problem and have introduced special grades of ultrapure, superpure, or electronic grade reagents and materials.[1-3] A list of suppliers of such products has recently been tabulated.[4] It is important to recognize that special nomenclature applied to a chemical is not significant; the results of a reliable analysis must define the purity of the material. A high purity chemical, purchased with or without a certified analysis, must be carefully characterized by a reference method to guarantee purity with respect to elements of interest.

Most suppliers of high purity chemicals furnish maximum values for trace constituents. One manufacturer has attempted to supply a growing list of inorganic and organic "ultrapure" chemicals with highly defined, actual-lot analysis.[5] An upper limit of 500 μg/g of impurity has been arbitrarily set for the ultrapure category.[6, 7] In the characterization of these materials the actual-lot analysis is insufficient; the details of the analytical method must also be provided on request.

An important concept in the commercial characterization of high purity chemicals is that a material must be equilibrated with its container before analysis. Accelerated aging studies of liquid chemicals in a variety of glass or plastic materials must precede the adoption of a suitable container. The difference between a reagent and an ultrapure chemical may, at times, reside in the container. For example, silicon tetrachloride with low iron content ($\sim$20 ng/g) can be prepared by careful fractional distillation. The iron content does not increase significantly when this product is distributed in a small (500 ml) borosilicate ampoule. Levels of 140–600 ng/g are found when larger quantities ($\sim$15 kg) are shipped in stainless steel drums.

For many analytical problems the level of a specific contaminant of interest can be adequately controlled only by designing a special laboratory purification procedure. Methods for the laboratory purification of reagents have been described in great detail by a number of workers.[8-13] The most important physical and chemical approaches to purity have also been critically reviewed.[14] A Russian monograph[15] also evaluated methods for preparing "superpure" inorganic chemicals containing less than 100 ng/g of impurities. This work contains many references to a steady stream of papers on ultrapurity published in *Trudy IREA* (*Proceedings of the All-Union Scientific Research Institute of Chemical Reagents and Ultrapure Chemicals*). Sixty references on investigations from 1948 to 1967 concerning ultrapurification by crystallization are cited. Unfortunately the *Proceedings* of the IREA are not abstracted by *Chemical Abstracts*.

The status of high purity chemicals in the USSR was reviewed in 1960.[16] Letters to the editor responding to this article were published later in the year [*Zavod. Lab.*, **26**, 1034 (1960)]. In general, complaints about the unavailability of ultrapure reagents for trace analysis were similar to the comments prevalent in Western countries.

This chapter describes state-of-the-art techniques for preparing ultrapure reagents in the laboratory. Special problems in the commercial production of high purity chemicals are also highlighted.

I. METHODS OF PURIFICATION

High purity reagents can be prepared in the laboratory by the methods listed in Table 1. In many cases classical techniques such as crystallization,

TABLE 1. Methods of Purification

Chromatography
  Partition
    Preparative gas-liquid chromatography
    Liquid
  Adsorption: adsorptive filtration
  Ion exchange
Crystallization
Distillation
  Fractional
  Isopiestic
  Subboiling
Electrolysis
Extraction
Fractional solidification
  Progressive freezing
  Zone melting
Ignition
Membrane filtration
Precipitation
Sublimation

fractional distillation, and liquid-liquid extraction are sufficient. Frequently one method can serve for prepurification and a second for ultrapurification. The success of these approaches frequently depends on appropriate starting materials.

Purification should be carried out in closed vessels by a process that ensures minimum handling of the product. Purification of sodium acetate, for example, is usually carried out in aqueous solution. The solvent is then removed to permit isolation of anhydrous crystals. Purification should be terminated at the solution stage whenever possible, because transfer of wet crystals and drying in ovens are prime sources of contamination. Chromium, nickel, and iron impurities in our early work were contributed by exposed heating elements in ovens. The contribution of airborne particulates via electrostatic effects during the transfer of a solid is an ever-present hazard. Inasmuch as ultrapure salts are invariably redissolved in water during trace analysis, the necessity for isolating the anhydrous salt must be evaluated at the inception of purification.

The washing of precipitates in open systems usually is a source of contamination. Figure 1 illustrates contamination control methods in a reaction vessel designed for precipitation, digestion, and washing of the precipitate. The vessel is constructed from a 2 gal polypropylene bottle. It consists of nylon, polyethylene, and polypropylene to protect the alkaline contents from contamination. An aqueous solution of magnesium nitrate, previously purified by extraction with 0.5 M thenoyltrifluoroacetone in 4-methyl-2-pentanone, is placed in the reaction vessel.[18] Hydrated magnesium carbonate is precipitated by the addition of ammonium carbonate with mechanical stirring at 90°C for several hours. The precipitate is vacuum filtered, then washed repeatedly with distilled water by back-filling through the three-way valve at the bottom. When the filtrate showed less than 2.5 μg/g of nitrate,[19] the washed precipitate was drawn off through the side shutoff valve into polypropylene cylinders for further processing. The dust cap on the stirrer is most important because most stirrers in the vertical position over an open system invariably deposit particulates into the reaction below. In our laboratory an improvement for the addition of liquids to the reaction has been introduced. A plastic fitting in the lid is connected to polypropylene tubing which, in turn, is connected to an all-Teflon (TFE) pump* for the addition of solutions at rates of 1–1000 ml/hr.

Monitoring of samples before and after purification by reliable analysis continues to be a major stumbling block. All too frequently the preparative chemist attempts to reduce metals to the nanogram per gram level when his analytical support is limited to the microgram range. As a result, the adop-

* Circle Seal Corp., P.O. Box 3666, Anaheim, Calif.

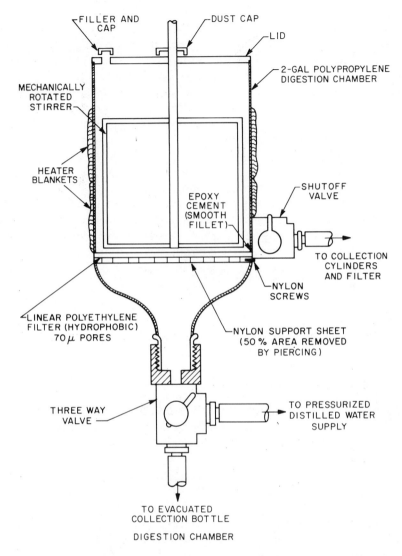

FILLER AND CAP

DUST CAP

LID

2-GAL POLYPROPYLENE DIGESTION CHAMBER

MECHANICALLY ROTATED STIRRER

HEATER BLANKETS

EPOXY CEMENT (SMOOTH FILLET)

SHUTOFF VALVE

TO COLLECTION CYLINDERS AND FILTER

NYLON SCREWS

LINEAR POLYETHYLENE FILTER (HYDROPHOBIC) 70 μ PORES

NYLON SUPPORT SHEET (50 % AREA REMOVED BY PIERCING)

THREE WAY VALVE

TO PRESSURIZED DISTILLED WATER SUPPLY

TO EVACUATED COLLECTION BOTTLE

DIGESTION CHAMBER

Fig. 1. Contamination control in a reaction vessel. Reprinted with permission from M. H. Leipold and T. H. Nielsen, *Am. Ceram. Soc. Bull.*, **45** (3), 281 (1966).

tion of the optimum purification scheme is endangered. For example, attempts were made in our laboratory to prepare sodium carbonate with iron, cobalt, and nickel controlled at the 1–5 ng/g level when the limit of detection for these elements was 10–50 ng/g. Aqueous solutions (20%) of sodium nitrate were extracted three times with chloroform solutions containing 8-quinolinol and ammonium pyrrolidine carbodithioate. The aqueous solution was passed through a column of an ion-exchange resin (strong sulfonic acid in H^+ form), and the resin loaded with the sodium was treated with ultrapure Na_2 EDTA [disodium salt of (ethylenedinitrilo)tetraacetate acid] to remove traces of di- and trivalent metals. The sodium was removed from the column by elution with high purity ammonium carbonate solution. When this purification sequence was repeated with improved detection limits for trace elements and thorough attention to air and container contamination control, it was noted with surprise that the extractions with the organic solution of complexing reagents afforded an aqueous solution that met the target specifications. The ion-exchange step is, therefore, superfluous. This type of "operation overkill" can be avoided by appropriate analytical support.

A. DISTILLATION

1. Acids

a. Subboiling Distillation

Subboiling distillation of HCl, HNO_3, $HClO_4$, H_2SO_4, and HF at the National Bureau of Standards has afforded extremely pure reagents.[20] Distillation of concentrated commercial acids batchwise in vitreous silica apparatus (Figure 2) was carried out in a laminar-flow hood.[21] The distillates were collected and stored in acid-washed, Teflon (FEP) 1000 ml bottles. The distillation of HF was carried out in a still of comparable dimension fabricated from polytetrafluoroethylene and warmed by electric resistance heaters inserted inside a glass tube, encased in a machined Teflon (TFE) rod. The stills were fed by a liquid level control that maintained the liquid at an appropriate height.[22]

Representative iron and lead analyses of acids and water purified in these subboiling stills are recorded in Table 2.[21] In all distillations except that of HCl the lead concentration is an order of magnitude lower than iron. In 10 N HCl the lead concentration is lower by a factor of 40. If a mineral acid is needed in a particular iron or lead trace determination, nitric acid provides the lowest blank. Sulfuric acid should be avoided whenever possible.

A novel, inexpensive subboiling distillation apparatus for the preparation of HCl, HF, and HNO_3 with ultralow lead content has been described.[23]

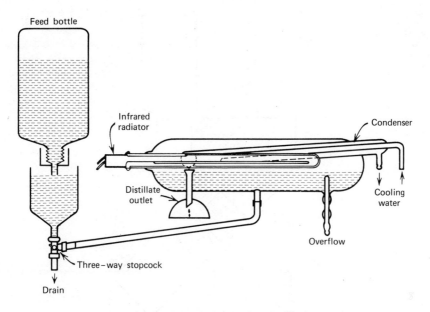

Fig. 2.   Subboiling still (vitreous silica).

The still (Figure 3) consists of two Teflon (FEP) 1000 ml bottles connected at right angles by a threaded Teflon (TFE) block. Heat is supplied by a 300 W heat lamp. The lead content of 48% HF was found to be 0.002–0.005 ng/g compared to 0.05 ng/g reported by the NBS investigators. The improved results can be ascribed to more efficient cleaning of an all-Teflon apparatus, minimum handling, and operation in a closed system. The Teflon apparatus is a refinement of the subboiling polyethylene still described by Coppola and Hughes.[24]

After standing for 2 weeks at room temperature, the iron content of three

TABLE 2.   **Iron and Lead Analyses of Acids and Water Purified by Subboiling**[a]

| Metal | \multicolumn{6}{c}{Amount of Metal (ng/g) in Water and Acids} |
	H_2O	HCl	HNO_3	$HClO_4$	H_2SO_4	HF
Iron	0.05	3	0.3	2	7	0.6
Lead	0.003	0.07	0.02	0.2	0.6	0.05

[a] Analyzed by stable isotope dilution mass spectrometry.
  Ref. 21.

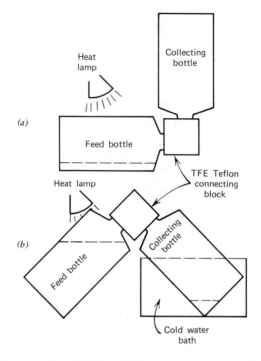

Fig. 3. Subboiling two-bottle still (Teflon FEP). (*a*) Cleaning position, (*b*) collecting position.

batches of 48% HF prepared in the all-Teflon apparatus averaged 5 ng/g compared to less than 1 ng/g for lead.[25] Since iron, zinc, aluminum, nickel, copper, and manganese impurities can be imbedded in the walls of all-Teflon (FEP) bottles (see Chapter 4), the relatively high iron content could result from the leaching of iron from the container walls. To prepare HF with ultralow transition element content, the Teflon (FEP) receiver must be decontaminated sufficiently or replaced by a polypropylene or high pressure polyethylene bottle.

b. Isopiestic Distillation

Isothermal distillation (also called isopiestic distillation) is a method for purifying volatile acids such as HCl, HBr, CH_3COOH, and HF.[26, 27] Actually, isothermal distillation is a minor variation of subboiling distillation. The following procedure for the purification of HF is a typical preparation. In a clean 30 × 20 × 10 cm Teflon (FEP) container equipped with a tight-fitting lid are placed two 250 ml polyethylene beakers approximately 5 cm apart. Technical grade HF (200 ml) is added to one beaker, and 200 ml of high purity water to the other. After the cover is snapped in place, the closed container is allowed to stand at room temperature. The water gradually absorbs pure HF vapor. To increase the concentration of the distillate, the original acid should be replenished every 2 days. After 4 days 75% HF yielded a product that assayed 50%. The lead content was reduced from 140 to 0.2 ng/g. The entire experiment should be carried out under a laminar-flow hood for optimum results.

Some investigators placed polyethylene vessels in a glass desiccator to prepare metal-free hydrochloric acid. Although plastic desiccators are available, they may be permeable to HCl vapors and must be placed in a hood. When the volume ratio of concentrated HCl to water was 2, 2 N HCl was obtained within 2 days.[26] When the ratio was increased to 10, 10 N HCl was isolated within 3 days. In one case a Vycor beaker (high silica, 96% SiO_2) containing 50 ml of water was immersed in 500 ml of concentrated HCl contained in a polyethylene pan covered with a polyethylene lid.[28]

c. Gaseous Saturation of High Purity Water

High purity water can be gassed with pure HCl, HBr, or HF to prepare acids of any suitable concentration. Figure 4 illustrates a standard

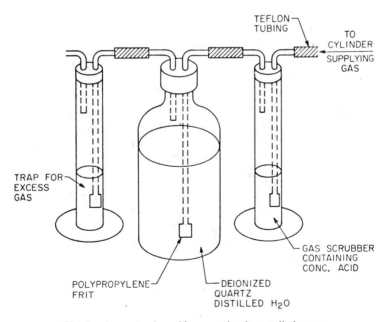

Fig. 4.   Apparatus for acid preparation from cylinder gas.

apparatus constructed from polyethylene, polypropylene, or polyfluorocarbon for the preparation of acids in this manner. An essential
component is a 0.45 μm Teflon filter at the outlet of the cylinder to remove
particulate matter. Tatsumoto[29] connected an HF commercial cylinder via
a filter to an all-Teflon train. The HF was liquefied by cooling, then
vaporized to remove volatile impurities. The purified HF fraction was then
added to high purity water. The lead content of aqueous HF was found to
be 0.08 ng/g. Of course the limitation of this method is the purity of the
water and the storage vessel at the end of the purification system.

A review of the laboratory methods for the preparation of high purity
volatile acids suggests that subboiling distillation in vitreous silica or Teflon
equipment provides the purest products. This technique has resolved the
handling, containment, and container-cleaning parameters with the greatest
success. The acids are usually stored in Teflon (FEP) bottles. Since
polyfluorocarbons withstand higher concentrations of acid at elevated
temperatures, impurities and inclusions can be removed from the interior
walls more readily during cleaning operations.

2. Bases

Isopiestic distillation of reagent grade ammonium hydroxide into high
purity water gives 10 N NH$_4$OH within 3 days.[26] Uptake of the ammonium
hydroxide by acidic samples rather than water is a good way to perform a
neutralization with a high purity reagent. If cylinder ammonia is first bub-
bled through an ammoniacal solution of EDTA, then through high purity
water with cooling, high quality aqueous ammonia is available. Gassing
water with ammonia directly from a cylinder is not recommended because
surprisingly high values for copper and nickel, 100 ng/g respectively, were
found in 10% NH$_4$OH solution prepared in this way. When impurities were
trapped by an EDTA solution, copper and nickel values were reduced to 4
and 1 ng/g.[30]

3. Solvents

Organic solvents are commonly purified in the laboratory by fractional
distillation. A 1.3 m borosilicate glass still (25 mm i.d.) packed with borosilicate helices and connected to a stillhead operating under total reflux can
upgrade reagent grade solvents from 99.0 to about 99.9% purity. The stillhead should contain a magnetically controlled takeoff for adjusting the
reflux ratio up to 100/1. When traces of boron are undesirable, the borosilicate should be replaced by vitreous silica.

Fractional distillation can be considered a prepurification step for several
solvents that can be further ultrapurified by fractional solidification (see

Fractional Solidification, Section I.C). Benzene, *p*-xylene, cyclohexane, and *p*-dioxane of 99.97% purity can be prepared in this way.

When organic solvents are employed to extract and preconcentrate trace metals from aqueous solutions, the metal blank of the solvent must be reduced. Reagent grade chloroform contains about 50 ng of lead per milliliter. After three distillations in a vitreous silica still and collection in a clean Teflon (FEP) bottle, the lead content was reduced to 0.012 ng/ml. When the reagent chloroform was first extracted with high purity 2*N* HCl before distillation, the lead content was further reduced to 0.002 ng/ml.

Oxygenated organic solvents such as ethers, alcohols, and esters have assumed increased importance in atomic absorption spectrometry. When neutral metal atoms are aspirated into a flame in the presence of these solvents, enhanced absorbancy results. Methylisobutyl ketone is preferred in many analytical laboratories. Distillation of this solvent in a vitreous silica still gives a product of satisfactory purity. Passage of MIBK through a column of silica before distillation removes acidic impurities.

Straight-chain or cyclic ethers (ethyl ether, 1,4-dioxane, and tetrahydrofuran) contain peroxides that must be removed before distillation. Peroxides are destroyed in ethyl ether by shaking with an acidic solution of ferrous sulfate and additional washing with water. The separated ether layer is then dried over sodium wire and distilled just before use. Acidification of dioxane, shaking with ferrous sulfate, and filtration are effective for this water-soluble material. For extraction spectrophotometry, distilled spectroquality solvents have been introduced that are characterized by maximum absorbance specifications in the ultraviolet region and assurance that the absorbance-wavelength curve is free from extraneous peaks. In addition, fractionally distilled high purity solvents provided with GLC assay, ultraviolet-visible-infrared absorbance curves, and conventional reagent analysis are available.[31] Perhaps the most dramatic assessments of compound control are found in the case of pesticide-free solvents. Here GLC coupled with an electron capture detector is sensitive to less than 1 part of heptachlorepoxide in 50 billion parts of solvents such as acetone, acetonitrile, and hexane.[32]

4. Water

Distillation removes water-miscible organics and ionizable inorganic impurities, as well as colloidal solids, from water. The quality of the resulting distillate varies with the original source of the water, the materials from which the distillation apparatus was constructed, and the number of distillations. Generally metal stills yield a product inferior to that obtained from borosilicate glass or vitreous silica (Table 3). Multiple distillations from vit-

TABLE 3.   Resistance of Distilled Water

Type of Distillation	Resistance $(M\Omega/cm)$
Single distillation in metal still	0.1–0.5
Single distillation in borosilicate glass	0.5
Three distillations in borosilicate glass	1.0
Three distillations in vitreous silica	2.0

Source: Ref. 33.

reous silica are usually required to obtain high purity water, which is also relatively free from organic traces. The addition of potassium permanganate to the stillpot has been a preferred technique for complete removal of organic contaminants, but it is not suitable for purification of water that is to be used for trace inorganic analysis.

The importance of the materials of construction was demonstrated on one occasion by replacing conventional borosilicate glass with polyethylene. When polyethylene, prepared by the high pressure process, was employed to condense and store the water distilled from a borosilicate still, the distillate had a resistance of 13 $M\Omega$. After storage for 2 weeks in polyethylene, the resistivity had dropped to 1.2 $M\Omega$, mainly because carbon dioxide had permeated through the walls of the container.[34] Other studies have shown that the trace element content of water stored in Teflon or polyethylene increased by a factor of 5 to 10 for some elements after storage for 30 days.[5] Distilled water, therefore, should be used as soon as possible after preparation. Cationic and anionic traces in water are effectively removed by ion-exchange resins. This technique is normally employed in sequence with distillation. The former treatment preceeds distillation because traces of organic amines are leached from anionic resins (see Ion Exchange, next section, for a comprehensive review on water purification).

B.   CHROMATOGRAPHY

1.   Liquid Chromatography

a.   *Ion Exchange*

Copolymers of styrene and divinylbenzene are the backbones of the resins most extensively used in the laboratory. A strong cation exchanger possesses $-SO_2OH$ groups; a strong anion resin contains $-NR_3OH$. When an aqueous salt solution is passed through a column packed with a mixed bed containing the H^+ form of the cation resin and the OH form of the anion resin, demineralization of the solution is accomplished rapidly and effectively.

The capacity of a strong cation resin (H^+ form) can be determined by passing a large excess of sodium chloride through a column of the resin, titrating the effluent with standard sodium hydroxide, and measuring the resin volume.[35] The capacities per milliliter of wet cation and anion strong resins are approximately 2 and 1.5 meq., respectively.[36] Therefore 500 and 700 ml of cation and anion resins will remove all of the Na^+ and Cl^- from 1 liter of 1 M NaCl solution. A mixed-bed resin is more effective than columns of cation and anion resins in series described in the early ion-exchange literature.[37]

The major ion-exchange materials employed in the laboratory are classified in Table 4. The sulfonic acid and quaternary ammonium functionalities receive extensive application in the removal of electrolytes from such water-soluble nonelectrolytes as carbohydrates, hexitols, glycerol, and formaldehyde.[38, 39] If a carbohydrate solution with high NaCl content is passed through a column of a sulfonic acid resin, the uptake of cations

TABLE 4. Chemical Structure of Ion Exchangers

Skeletal structure
  Polystyrene-divinylbenzene copolymer
  Phenolic
  Acrylic
  Cellulose
  Dextran
Functional groups
  Cation exchangers
    Sulfonic acid RSO_2OH
    Carboxylic acid $RCOOH$
  Anion exchangers
    Quaternary $RCH_2N(CH_3)_3Cl$
    Secondary amine $RR'NH$
    Tertiary amine $RN(CH_3)_2$
  Chelating resins

$$\text{Iminodiacetic acid} \qquad R-N\begin{cases} CH_2COOH \\ CH_2COOH \end{cases}$$

$$\text{Amidoxime} \qquad R-C\begin{cases} NOH \\ NH_2 \end{cases}$$

$$\text{Phosphonic} \qquad R-P\!\!\overset{\displaystyle O}{\underset{\displaystyle OH}{\text{---}OH}}$$

produces an acid eluate that may attack the substance undergoing purification. The use of a mixed-bed resin, on the other hand, assures the prompt removal of acid and base from a solution and produces a neutral eluate instantly.

A mixed bed of strong cation (sulfonic acid) and weak anion (tertiary amine) resin removes all cations and strong acids such as HCl, H_2SO_4, HNO_3, H_3PO_4. In this system acids such as carbonic acid, boric acid, silicic acid, and weak organic acids are not adsorbed by the tertiary amine groups and can, therefore, be demineralized with ease. A weak carboxylic acid cation exchanger in the sodium form is unusually selective for removal of copper, iron, and nickel at pH values above 4.

A number of chelating ion-exchange resins for the separation of multivalent metal ions from monobasic cations have been synthesized.[40] Chelating resins resemble the monomeric chelating agents in that the same order of stability constants with metal ions is retained. Traces of metals that form chelates can be removed from a large excess of Na, K, NH_4, or Li salt solutions by percolating the solution through a column of chelating resin. Derivatives of iminodiacetic acid have received frequent attention. Dowex chelating Resin A-1,[41] for example, is a copolymer of styrene, divinyl benzene, and vinylbenzylamino diacetic acid. The last monomer is prepared by direct alkylation of iminodiacetic acid with vinylbenzyl chloride.

Alkali and alkaline earth salts can be freed from transition elements by passage through columns of amidoxime resins. These resins are pepared by treating polyacrylonitrile with hydroxylamine hydrochloride.[42] Very strong chelates are formed on the resin with Fe^{3+}, Cu^{2+}, $V^{2, 3, 4, 5+}$, $Pt^{2, 4+}$; moderate chelation with Zn^{2+}, Cd^{2+}, $Cr^{2, 3+}$, $Ni^{2, 3+}$; weak chelation with $Mn^{2, 3+}$, $Pb^{2, 4+}$.

An effective application of this resin in ultrapurification was demonstrated by the removal of iron from calcium nitrate solutions. A 2 M solution of calcium nitrate (Fe = 3 ppm) is filtered through a fine cellulose paper of about 2 μm porosity. The filtrate (Fe = 0.3 ppm) is then stirred overnight at room temperature with amidoxime resin beads* (weight of resin = 0.5% of weight of calcium nitrate). The iron content in the eluate is 0.01 ppm as measured by ferrozine photometry.[43]

When sulfonic acid groups in conventional strong cation resins are replaced by phosphonic acid $-\overset{\displaystyle O}{\underset{\displaystyle OH}{P}}-OH$, a resin is formed with unusual affinity for removal of lead, copper, zinc, iron, cadmium, or manganese.[44]

---

* Duolite CS-346, Diamond Shamrock Chemical Co., Redwood City, Calif.

The advantage of the resin is dramatically indicated by the measured relative affinity for ions (Table 5). Alkali salt solutions percolating through this resin are separated from transition metals or other metals forming insoluble phosphates. Examination of tables presenting the chemical and physical properties of commercial resins aids immensely in the selection of the proper resin for a particular problem[45] (see Appendix).

(1) *Preparation and Purification of an Ion-Exchange Column.* A sulfonic acid resin of U.S. mesh size (wet screen analysis) 16-50 and 8% cross-linkage (8 parts divinylbenzene, 92 parts styrene) is suitable for most applications. The 100-400 mesh is reserved for chromatographic fractionation. Commercial resins are produced in 4, 8, 12, and 16% cross-linkage. For monovalent cations the relative affinity, based on hydrogen ion as 1.0, varies from 1.3 to 3.5 in proceeding from 4 to 16%. Copper and silver, however, increase from 3 to 17. Divalent cations increase from 2 to 8 in most cases. Mercury, lead, and barium increase from 5 to 16.5.

The resin is available in the H^+ or Na^+ form. Maximum specifications for copper, iron, and lead impurities are approximately 50, 500, and 10 ppm. A typical lot analysis is closer to 1, 30, and 1 ppm, respectively. The sodium form of the resin should be added to a 50 cm height of high purity water in a polymethyl methacrylate, polyethylene, or polypropylene tube (25 mm $\times$ 75 cm) provided with a polyethylene frit at the bottom to support the resin and a polyethylene stopcock to control the flow of eluate. The resin is purified by eluting with 6 column volumes of a dilute solution of Na_2 EDTA [the disodium salt of (ethylenedinitrilo)tetraacetic acid, 4 g/liter adjusted to pH 7.0] followed by 6 column volumes of the same solution adjusted to pH 9.0. The resin is finally washed with 6 bed volumes of high purity water. Since the higher oxidation states of metals enhance their removal from resins by EDTA, cation resin (sodium form) should first be eluted with large volumes of 1 (1 M NaOH):7 (30% H_2O_2). This mixture removes Cr^{3+} from the resin by oxidizing it to the anion CrO_4^{2-}. Any Fe^{2+} and Co^{2+} on the resin is simultaneously oxidized to Fe^{3+} and Co^{3+} before sub-

TABLE 5. Relative Affinity[a] of Resins in 0.05 M Solutions

Resin	Na^+	Ba^{2+}	Sr^{2+}	Mg^{2+}	Ca^{2+}	Ni^{2+}	Co^{2+}	Mn^{2+}	Cd^{2+}	Zn^{2+}	Cu^{2+}	Pb^{2+}
Duolite ES-63[b]	0.2	2.0	2.0	2.3	3.0	17	23	51	195	370	890	5000
Dowex-50, IR-120, C-20[c]	1.5	8.7	4.9	2.5	3.9	3.0	2.8	2.3	2.9	2.7	2.9	7.5

[a] All values are compared with the affinity of strong cation resins for hydrogen (affinity = 1.0).
[b] Phosphonic Chelating Resin, Diamond-Shamrock Co.
[c] Dow, Rohm and Haas, and Diamond-Shamrock strong cation (sulfonic acid type).

sequent elution with EDTA solution at pH 9.0. Vanadium can be removed from the acid form of the resin as a vanadium peroxide complex by elution with dilute acid containing 1% H_2O_2.

Efficient removal of cations from ion-exchange resins by elution with EDTA was evaluated with radiotracers as follows. A 2 ml volume of solution containing 2 μg of Fe^{3+}, Co^{2+}, Mn^{2+}, and Cr^{3+} was doped with enough tracer to produce approximately 10^5 cpm for each of the isotopes ^{59}Fe, ^{60}Co, ^{54}Mn, and ^{51}Cr. Because of the high specific activity of the tracers, no changes in the total concentration of each element occurred. The solution was then passed through a 15 cm column of the sodium form of Amberlite IR-120 cation resin in a 9 mm $\times$ 22.5 cm polyethylene tube. After the exchange of the ions in the tracer solution, the column was eluted with at least 5 column volumes of deionized water. The absence of γ-ray activity in the fractions collected with a Beckman Model 132 automatic fraction collector indicated complete exchange of the ions in the tracer solution. The ion-exchange resin was then purified by elution with 700 ml of a solution containing 4 g/liter of Na_2 EDTA adjusted to pH 9.0. Ten milliliter fractions of the eluate were collected, and 5 ml aliquots were counted in a well-type NaI detector attached to a TMC Model 1001 multichannel analyzer. The γ-ray activity remaining on the resin was measured by transferring the entire amount of resin to vials and counting in the manner previously described. The γ-ray spectrum produced by isotopes that remained on the resin after elution with Na_2 EDTA shows the incomplete removal of Cr^{3+} (Figure 5).[46]

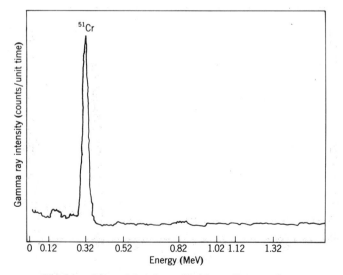

Fig. 5. γ-Ray activity of a purified ion-exchange resin.

The efficiency of the procedure for purification of the ion-exchange resin was determined by counting γ-ray intensities of the individual elements at characteristic energies and by comparing the initial activity of the elements on the resin with the activity after elution with Na_2 EDTA. After elution with 175 column volumes of Na_2 EDTA solution, 76%, < 0.001%, and 0.02% of the initial amounts of Cr^{3+}, Mn^{2+}, and (Co^{2+} + Fe^{3+}), respectively, remained on the resin.

A solution containing 2 μg of Cu^{2+} and VO_2^+ was spiked with ^{64}Cu + $^{48}VO_2^+$ and eluted through the Na^+ form of the resin as before. The $^{48}VO_2$, absorbed on the resin, produced a γ-ray intensity of 4.4×10^4 cpm. A total activity of 8.7×10^4 cpm was generated by the mixture of $^{64}Cu^{2+}$ and $^{48}VO_2^+$. After elution with 300 ml of Na_2 EDTA solution at pH 9.0, 40% of the original amount of VO_2^+ remained on the resin. The γ-ray spectrum of the isotopes on the resin, compared with a known spectrum of $^{48}VO_2^+$ and that produced by a mixture of $^{48}VO_2^+$ and $^{64}Cu^{2+}$, showed that copper was completely removed from the resin. To remove Cr^{3+} and VO_2^+, the sodium form of the sulfonic acid is purified first by elution with 50 column volumes of 1 (1 M NaOH):7 (30% H_2O_2).

The sodium is removed by eluting the column with 5 M HCl, then with water, until the eluate is neutral to pH paper. A purified sulfonic acid resin in the H^+ form is then available.

(2) *Water Purification.* The major application of mixed-bed resins is the purification of water, a primary concern to trace analytical chemists. This major-volume reagent must be monitored at regular intervals because its purity can support or undermine an entire analytical operation. The ACS requirements for a standard quality of water, reagent grade, are listed in Table 6. Although a test for specific conductance (S.C.) is described in the specifications, measurements of specific resistance 1/S.C. are usually simpler to perform on an in-line direct-reading resistivity meter.

TABLE 6. American Chemical Society Specification for Water, Reagent Grade

Specific resistance at 25°C	>0.5 MΩ/cm
Silicate (as SiO_2)	Not more than 10 ng/ml
Heavy metals (as Pb)	Not more than 10 ng/ml
Substances reducing permanganate	To pass test[a]

[a] To 500 ml add 1 ml of sulfuric acid and 0.03 ml of 0.1 N permanganate and allow to stand for 1 hr at room temperature. The pink color should not be entirely discharged.

Source: Reagent Chemicals: American Chemical Society Specifications, 5th ed., American Chemical Society Publications, Washington, D.C., 1974, p. 663.

TABLE 7.   **Specifications for Reagent Grade Water**

Property Specified	New Specifications			
	Type I	Type II	Type III	Type IV
Maximum total matter (mg/l)	0.1	0.1	1.0	2.0
Maximum conductivity (μmho/cm at 25°C)	0.06	1.0	1.0	5.0
Minimum resistivity (MΩ/cm at 25°C)	16.66	1.0	1.0	0.20
pH at 25°C	6.8–7.2	6.6–7.2	6.5–7.5	5.0–8.0
Minimum color retention time of KMnO$_4$[a] (min)	60	60	10	10

[a] Current test specifies the addition of 0.40 ml of 0.01 N KMnO$_4$ per liter. In the original specifications 1.5 times this amount was added.

Revised ASTM specifications for four grades of reagent water appear in Table 7.[47] Type I, "ultrapure," water is recommended for the preparation of solutions for trace metal analysis. Type II water, usually obtained by double distillation, is satisfactory when freedom from organic material is more important than freedom from trace metal ions. Type III water is satisfactory for preliminary washing and rinsing of laboratory glassware in the analytical laboratory. Type IV is useful where maximum purity is not essential. It serves adequately when large quantities of water are needed to flush degradation products from ion-exchange resins or to prepare laboratory reagents in which trace impurities can be tolerated.

The specifications for reagent water reflect the classical method of distillation developed to eliminate "organics" such as pathogens and pyrogens in water destined for parenteral use in the pharmaceutical industry. Improvements in the design and performance of all-glass laboratory stills have recently been reviewed.[48, 49] A borosilicate glass still operating on tap water can now furnish a pyrogen-free distillate with a resistance of 1.8 MΩ/cm; the resistance increases to 8 MΩ/cm when the tap water is demineralized through a strong base–strong acid–mixed-bed resin prior to distillation.*

Resistance measurements give only qualitative indications of the presence of dissociated ions. Particulate matter, nonionized materials, and charged molecules with low solution mobilities are not detected. Table 8 illustrates qualitatively the levels of electrolyte content reported for increasing specific resistance of water.[35] Table 9 reports the actual cation contamination of four grades of water. Tap water, as expected, was found to be high in calcium, magnesium, and sodium. Upon distillation in a commercial metal still, substantial improvement was noted, but the still contributed copper

* Ultrascience, Inc., Water Purification Division, Evanston, Ill.

TABLE 8. Electrolyte Content of Water

Specific Resistance (MΩ/cm at 25°C)	Approximate Electrolyte Content (ng/g)		
	NaCl	HCl	CO_2
0.1	4000	1500	70,000
0.5	800	260	2,500
1.0	400	130	800
10.0	40	10	

and lead.[50] Purification of the tap water with a train of carbon, mixed-bed resins, and a 0.2 μm cellulose acetate filter in an all-plastic apparatus* gave an effluent suitable for sophisticated trace analyses. When this water was substituted for distilled water (metal still) purified by mixed-bed resins and a 0.45 μm cellulose acetate filter, lower blanks were found in a variety of ultratrace analyses. Water with the lowest cation content was obtained by

TABLE 9. Analysis of Water for Cations (ng/g)

Element	Water Samples			
	A[a]	B[a]	C[a]	D[a, b]
Ag	<1[51]	1.0[50]	0.01[52]	0.002[20]
Ca	>10,000	50.0	1.0	0.08
Cd	—	—	<0.1	0.005
Cr	40	—	<0.1	0.02
Cu	30	50.0	0.2	0.01
Fe	200	0.1	0.2	0.05
Mg	8,000	8.0	0.3	0.09
Na	10,000	1.0	1.0	0.06
Ni	<10	1.0	<0.1	0.02
Pb	<10	50.0	0.1	0.003
Sn	<10	5.0	<0.1	0.02
Ti	10	—	<0.1	0.01
Zn	100	10.0	<0.1	0.04

[a] Sample A, tap water, Phillipsburg, N.J.; sample B, tap water purified by two-stage distillation (commercial metal still); sample C, tap water purified by train of carbon, mixed-bed resins and 0.2 μm cellulose acetate filter; sample D, deionized water purified by subboiling distillation.
[b] Sample D was analyzed by isotope dilution; samples A, B, and C by emission spectrography after concentration of a 1000 ml sample.

* Super-Q System, Bulletin MB-403 (1973), Millipore Corp., Bedford, Mass.

subboiling distillation in a vitreous silica still operating in a vertical laminar-flow hood.[21]

In one commercial vitreous silica still* (Figure 2) the water is warmed by an electrically heated infrared radiator enclosed in a vitreous silica sheath placed above the surface of the liquid. The water surface always remains quiescent; thus no liquid can creep along the walls of the equipment. Apparatus with hourly outputs of 200, 500, and 1500 ml is available. The sterile nonpyrogenic distillate is ideally suited for analyses in the biological laboratory as well as in trace analytical work.

High purity water can easily be contaminated by the vessel in which it is contained. Once in our laboratories water (sample C in Table 9) was monitored by withdrawing 3 liters into a polyethylene bottle previously cleaned with nitric acid and thoroughly rinsed with high purity water. After the evaporation of 1000 ml portionwise in a Teflon dish, emission spectrographic analysis of the residue showed surprisingly high calcium values (7 ng/g). Investigation of the history of the polyethylene container revealed that high purity solutions of 1 M $CaNO_3$ had been stored in the bottle for 2 months prior to cleaning with nitric acid. The calcium content of the water decreased to 1 ng/g upon withdrawing the water into a new polyethylene bottle that had been subjected to the standard cleaning procedure. High sodium values (8 ng/g) were found when polyethylene containers previously used for storage of 25% sodium carbonate were cleaned and used for water sampling.

Protection from particulate contamination is a primary concern after dissociated electrolytes and organics are removed from water. The particulate matter content of five water samples is given in Table 10. The most striking observation is that water contains 5 and 10 μm particles after filtration through a 0.2 μm cellulose acetate filter. This finding has been checked by several pharmaceutical laboratories for all membrane filters commercially available. With the shift to a 0.45 μm filter, the number of 5 μm particles in the filtrate increases. Some laboratories use cartridges of mixed-bed resins without proper filtration of the effluent water. Resins are not designed for efficient particulate matter removal; in fact, fine resin particles usually separate from the resin column. Also the increase in 5 μm particles from 12 to 6350 per 10 ml by inelegant finger-stirring (Table 10) demonstrates that the hands are, indeed, a prime source of particulate contamination. A unit containing noncellulosic membranes with pore diameters in the range 0.001–0.02 μm has recently been introduced. These hollow fiber cartridges† assist in the complete removal of suspended and colloidal particles from ion-exchanged or distilled water.

* Quartz Products Corp., Plainfield, N.J.
† Romicon, Inc., 100 Cummings Park, Woburn, Mass.

TABLE 10. Particulate Matter (particles/10 ml) in Water[a]

Method of Purification	Size of Particles (μm)			
	5	10	15	20
A. Tap water	4180	800	263	109
B. Tap water (purified with carbon, mixed-bed resin, and 0.2 μm cellulose acetate filter)	12	5	1	0
B stirred with magnetic bar	39	12	5	2
B stirred with forefinger	6350	1002	337	75
B plus distillation (0.45 in place of 0.2 μm filter)	45	12	4	3

[a] Analyzed by PC-305 unit, High Accuracy Products Corp., Claremont, Calif.

Analysts frequently have water transported from a central distillation or deionization system into the analytical laboratory. Personal experience, however, has shown that filters, ion-exchange resin, or carbon beds may not be properly maintained in community-type systems. Furthermore, water obtained through improperly constructed distribution lines has added 50 to 100 ng of lead per milliliter to effluent water when the source water from public supplies contained less than 5 ng of lead per milliliter. If the complete reliability of a central system is not controlled by the trace analyst, he should not jeopardize his results by faithfully using the distilled deionized water faucet conveniently located above his sink. Instead, he should install a dedicated facility, ensuring that high purity water is available preferably within a laminar-flow hood, thus eliminating airborne contamination at the point of use.

In the system used in one trace analytical laboratory (Figure 6), distilled water is fed at a flow rate not exceeding 100 ml/min through tube A into a series of mixed-bed ion exchange columns B_1 and B_2. The ion-exchange system is composed of two plastic columns with all-Teflon fittings connected by Teflon tubing. After deionization, the water passes through a 0.2 μm Teflon filter C to remove particulates; then it flows into the lower reservoir of the all-quartz still E. Upon distilling from the lower chamber and condensing in the upper chamber F, the water is redistilled, condensed, and passed from the collecting port H into Teflon tubing I, which leads to a polypropylene holding tank. Data obtained by a semiquantitative spark source mass spectrometric analysis of water produced with this system are reported in Table 11.[54] By directly comparing samples of purified water with equal volumes of water doped with known amounts of trace elements,

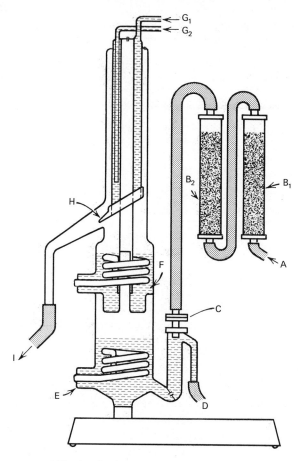

Fig. 6. System for purification of water.

the maximum concentration of each element was determined to be less than 0.1 ng/g.

Water purified by this procedure should be stored for a maximum period of one week before use in ultratrace analysis. Unused water is used in less demanding applications or discarded. Storage of purified water, even in polypropylene or polyfluorocarbon containers for periods exceeding 30 days, will result in an increase in cationic impurities.[54, 55] Purified water shows much greater concentrations of aluminum, copper, iron, lead, and zinc when stored in borosilicate glass bottles rather than polyethylene or polypropylene.[13]

Another popular system for water purification depends basically on an ion-exchange method. Tap water is prepurified by passage through a filter

TABLE 11.   Mass Spectrometric Determination
of Trace Impurities in High Purity Water[a]

Element Detected	Concentration in Sample (g/ml)	Blank[b]
^{55}Mn	$\sim 1 \times 10^{-10}$	N.D.
^{58}Ni	$\sim 1.2 \times 10^{-10}$	N.D.
^{63}Cu	$< 1 \times 10^{-9}$	N.D.
^{51}V	1×10^{-10}	N.D.
^{59}Co	$< 1 \times 10^{-10}$	N.D.
^{52}Cr	$< 1.2 \times 10^{-10}$	N.D.
^{64}Zn	5×10^{-10}	M+
^{32}S	1×10^{-10}	M+
^{35}Cl	1.2×10^{-10}	—
^{56}Fe	1×10^{-10}	M+
^{40}Ca	1×10^{-10}	—
^{39}K	1×10^{-10}	—

[a] 0.02 ml of sample and equal volumes of doped standard were evaporated on Ta electrodes (see Chapter 7, Spark Source Mass Spectrometry, Section V, for procedures).
[b] M+ = monovalent ion faintly detected; N.D. = not detected.

and two mixed-bed resin columns. Under a service contract,* the filter and the first column are replaced regularly when exhausted. The prepurified water is then passed through a train of carbon, mixed-bed resins, and a 0.2 μm membrane filter to give the results reported for sample C in Table 9. Prepurification of tap water with the first two mixed-bed resin columns prolongs the life of the more expensive carbon, resin, and polishing filter of the Super-Q System. Effluent water is available in a class 100 clean area at the rate of 1100 ml/min. Thus no storage facilities for purified water are required.

Water can also be prepurified before distillation or deionization by reverse osmosis units that remove 95% of dissolved inorganic salts, considerable organic material, and particulate matter via a semipermeable membrane and a 5 μm filter. Reverse osmosis equipment is worthwhile if a laboratory uses more than 100 gal of water daily. This water costs as little as one-tenth of a cent per gallon and is adequate (Type III) for dishwashing in the laboratory.

(3) *Hydrochloric Acid and Sodium Hydroxide.*   Although strong anion, quaternary ammonium resins normally remove strong and weak acids from

* Continental Water Conditioning Corp., P.O. Box 26428, El Paso, Texas.

solution, anionic complexes of cations can also be adsorbed. For example, Fe^{3+}, Sb^V, Ga, and Au form strong complexes with 8–12 N HCl.[56] The complexation reaction for Fe is typical for these metals,

$$FeCl_3 + HCl \rightleftharpoons H^+ + FeCl_4^-$$

The $FeCl_4^-$ is bound more firmly than Cl^- on the chloride form of a quaternary ammonium resin. The distribution constant is 1000 for the $FeCl_4^-$ complex. Consequently, Fe^{3+}, Sb, Ga, and Au can be removed from concentrated HCl by simply percolating the acid through a column of the resin. As 1.3 meq. of the complex can be adsorbed by 1 ml of wet resin (IRA-400, Dowex 1-X8), theoretically 72.8 mg of iron can be removed from concentrated HCl by this small quantity of exchanger. Inasmuch as reagent grade HCl contains a maximum of 0.1 ppm Fe, small columns of resin (1 × 10 cm) can remove ferric iron from hundreds of liters of concentrated HCl. Ferrous iron must be removed by distillation.

Sodium hydroxide solutions are more difficult to purify than HCl. Trace iron exists as a cation in dilute NaOH solutions but as sodium ferrate ($NaFeO_2$) in concentrated solution. Sodium hydroxide solutions, therefore, should be passed through a mixed bed of strong cation (a sulfonic acid in the sodium form) and strong anion (quaternary ammonium in OH form) resins. In concentrated solutions of sodium hydroxide, the resin is decomposed; the eluate is then colored by degradation products from the resin.

b. Adsorptive Filtration

A single filtration over an active adsorbant such as alumina or silica can be used to purify organic solvents sufficiently for spectrophotometric applications. The removal of polar impurities such as peroxides, water, alcohols, and acids from aliphatic hydrocarbons[57, 58] and carbon tetrachloride or dimethyl sulfoxide[59] can be carried out very simply. The most effective column materials are alumina and silica with activity I, as defined by Brockmann.[60] The activity of these oxides depends on the water content and surface area. Since an increase in water content lowers the activity, adsorbants packaged in water-permeable plastic containers must be tested before use as follows.

In a dry 10 ml test tube suspend 2 g of the adsorbant in 3 ml of a 1% solution of triphenylchloromethane in dry benzene.[68] A yellow to light-brown color indicates activity I; a colorless solution defines an adsorbant with low activity (> 0.3% water).

To determine Grades I–V according to Brockmann, five solutions are prepared containing 2 mg of one of the following dyes in 10 ml of benzene-petroleum ether (1:4):p-methoxyazobenzene, Sudan yellow, Sudan red, aminoazobenzene, and hydroxyazobenzene. Each solution is then added to a 50 mm length of alumina packed in a 15 × 100 mm borosilicate tube. The column is then developed with 20 ml of benzene-petroleum ether (1:4). When the following individual dyes remain in the top 10 mm of the column, the grade is indicated within the adjacent parentheses: p-methoxyazobenzene (I), Sudan yellow (II), Sudan red (III), aminoazobenzene (IV), hydroxyazobenzene (V). Adsorbants can be activated by removal of water at 200°C in an oven under a nitrogen atmosphere. A given weight of silica is more efficient than alumina because of the 3–5 times greater surface area of the silica. The purification of n-hexane is a typical example of the utility of this method. A one-piece borosilicate chromatography tube (15 × 450 mm), equipped with a 2 mm bore Teflon stopcock and a coarse sintered glass filter disk to support the column packing, is filled with silica, activity I. Reagent grade n-hexane is passed through the column at a rate of 20 drops/min. Usually 80 g of silica is sufficient to purify 500 ml of n-hexane (80 g of alumina purifies only 100 ml). Examination of the absorption curves of the starting material and the purified fraction of the eluate showed a shift from 273 to 218 nm at 50% transmission.

c. High Performance Liquid Chromatography

The separation efficiency of conventional ion-exchange, absorptive, and partition column chromatography has been notably improved by experimental advances in the past 10 years.[62] The commercial availability of 5–50 μm microparticles has facilitated complex separations. It is not unusual for a 10 μm silica column packing to have 32,000 plates per meter.[63] The only disadvantage of these high efficiency column packings is that high pressure is required to force the mobile phase through the column.

Commercial equipment for high performance liquid chromatography (HPLC) is now common in most analytical laboratories. Analytical columns have an internal diameter of 2–2.6 cm. However a recent study has described the performance of a preparative HPLC unit with the column internal diameter increased from 2 to 15.8 cm.[64] The sample undergoing purification was increased from 20 to 875 μg. At a pressure of 7 atm the sample was purified in a 41 cm length of column in 28 min. The chromatograms for separations on the analytical and preparative columns were strikingly similar, even though the efficiency for the preparative column (~ 1000 plates per meter) was twice that for the analytical column.

Preparative HPLC is an excellent technique for the isolation of high purity, nonvolatile, thermally unstable, organic standards. Once the conditions for a purification have been worked out, repetitive runs in an HPLC instrument can yield milligrams of a standard during an 8 hr period.

2. Gas-Liquid Chromatography

Gas-liquid chromatography became an indispensable analytical tool from the moment James and Martin announced the technique in 1952.[65] In 1957 Kirkland described the first laboratory-scale apparatus for the separation of kilogram quantities of ultrapure liquids by GLC. In the past 19 years commercial instruments have become available for scaling up analytical separations. Any separation obtained on a microliter scale that shows ultrapurification at 99.95% or greater can be scaled up with the appropriate liquid stationary phases to give kilogram quantities.

When the stationary liquid phases are stable at elevated temperatures, some columns can be operated at 400°C. In most applications the boiling points of liquids undergoing purification are below 200°C at 760 mm. For example, the physical constants of isopropylbenzene (b.p. 152°C; m.p. −96°C) immediately suggest ultrapurification by GLC. Actually this material was purified on a 2 liter scale by preparative GLC.[67] A commercial product (98.72% purity) was volatilized on a column (length, 200 cm; diameter, 10 cm) packed with 5 gal of 20% Carbowax M on 10-60 Chromosorb P. A series of 50 ml volumes was volatilized onto the column at a rate of 3.5 ml/sec; a manifold collection valve automatically collected the ultrapure (99.97%) fraction only. Organic materials are assigned the ultrapurity designation when the assay is 99.95% or greater.

Reagents, solvents, or standards, purified by preparative GLC, often need to be redistilled in glass to eliminate contamination from bleeding of the stationary liquid phase. Although the bulk of the literature stresses GLC application to organics, some inorganic compounds are also readily chromatographed.[68] Metal chlorides and chelates that have been separated on analytical columns can also be isolated on a preparative scale.

C.   FRACTIONAL SOLIDIFICATION

Solids that melt without decomposition can be purified by the unequal distribution of impurities at the solid-liquid interface. For the most part impurities remain in the melt during the crystallization of the host chemical. The purification of benzoic acid, for example, by the simple process of directionally solidifying a tube of molten acid while stirring the contents, increased the purity from 99.91 to 99.9997 mole %.[69] After 30 years, renewed interest has developed in the application of this procedure to laboratory-scale purification.[70] Progressive (also called directional or fractional) freezing is successful for upgrading material of 90–95% purity. Zone melting[71, 72] and column crystallization[73, 74] require starting material at least 98.5% pure. Although the last two methods are used to attain the highest purity, fractional freezing is satisfactory when a 99.95% pure product is the objective.

1. Progressive Freezing

Preparation of ultrapure dioxane ($\geq$ 99.95% pure) involves pre- and ultrapurification. To prepurify, prepare 99.9% material by fractional distillation (see Distillation, Section I.A).

Ultrapurification is begun by placing 1 liter of prepurified dioxane in a 2 liter, round-bottomed borosilicate flask. Stir with an all-glass assembly as the flask is slowly cooled to 5°C in an isopropanol cooling bath controlled by a low temperature circulating unit.* As the dioxane freezes, raise the stirrer, keeping the blade always in the liquid portion. After 90% of the charge is frozen, siphon off the supernatant and discard. Remove the cooling bath, thus allowing the solid to melt. Repeat the freezing step and again separate the supernatant (10% of the contents). The solid ($\sim$ 720 g) is then remelted and analyzed by GLC. Usually five or six peaks in the prepurified material disappear; the product is characterized by one peak only (minimum purity, 99.99%).

Fractional freezing provides inexpensive, rapid ultrapurification for the following organics; the melting point (°C) appears in parentheses: cyclohexane (6.5); acetic acid (16.6); cyclohexanone (-16); dioxane (12); aniline (-6); carbon tetrachloride (-23); p-xylene (14). *The boiling points (144, 139, and 139°C) and the melting points (-25, -47, and $+13°C$) of the ortho-, meta-, and para-isomeric xylenes clearly demonstrate that freezing should be a powerful purification tool.* Actually p-xylene (99.99% assay by GLC chromatography) was obtained after partially freezing 99.86% distilled material.[75]

2. Zone Melting

Whereas the removal of the supernatant in each cycle of progressive freezing confines this procedure to a batch process, zone melting can be completely automated. A molten zone is passed repeatedly through an ingot to effect purification by continuous crystallization from a melt. A valuable piece of equipment for ultrapurification is a vertical zone melter† suitable for the purification of organic and inorganic compounds melting from -10 to 300°C.[76] Salts (NaCl, m.p. 801°C) and metals (Fe, m.p. 1535°C) can be zone refined,[77, 78] but purification at these temperatures is not recommended for the average laboratory.

* Model TK-30, Lauda Table Model Kryomat, Brinkman Instruments.
† Sloan-McGowan Zone Melter, Princeton Organics, Princeton, N.J.; Desaga DBGM Zone Refiner, Camlab, Cambridge, England.

In the ultrapurification of benzoic acid, a primary standard, first clean a borosilicate sample tube 8 mm o.d. and 50 cm long with hot nitric acid, then rinse with high purity water and allow to dry in a laminar-flow hood. Fill the tube with reagent grade benzoic acid to within 6 cm of the top, seal the tube with a hand torch, and fasten a glass hook to the top for attachment to a drive pulley. The open space above the free surface of the solid is very important because a large increase in volume accompanies fusion for many solids. If a molten zone cannot form at a free surface, the container will break. To allow for expansion, the top of the tube is first suspended below a battery of nine heaters spaced at intervals of 5 cm, then drawn up slowly at a rate of 25 mm/hr until the entire tube is within the series of heaters. The tube is lowered through the distance corresponding to 5 cm spacing between heaters and is finally controlled automatically so that it rises through one interzone distance (5 cm) at 25 mm/hr and descends the same distance quickly (0.83 cm/sec). The cycle of 9 heating zones per pass can be continued automatically from 1 to 100 passes. After the desired number of passes has been completed, the tube is cooled, detached from the pulley, wrapped in a polyethylene sheet, and fractured. The top and bottom 5 cm sections of the ingot are discarded. Discrete lengths ($\sim$ 2 cm) of the remaining ingot are then analyzed. Differential scanning calorimetry measurements are reliable up to 99.99% purity.[79] Beyond this value, assay of benzoic acid by a routine method is difficult. High precision coulometry must be used.

Benzoic acid has been purified at zone velocities of 8.3–93 mm/hr after 5–274 passes. Zone lengths have varied from 3 to 12 mm in tubes 2.5–22 mm in diameter and 10–60 cm long.[79-85]

Zone-refined products are usually isolated as solid chunks. When Teflon tubing with glass plugs at the ends replaces the original sample tube, the chunks are removed more easily after purification.[86] Shattering of glass is avoided because the tube can be sliced with a razor blade. Ultrapure fractions are then combined and sublimed to obtain free-flowing powders.

3. Column Crystallization

Continuous ultrapurification on a laboratory or industrial scale has developed from the work of Schildknecht and Vetter,[87] who placed a rotating spiral in the annulus between vertical, concentric tubes. The material to be purified occupies the annular space between the two stationary tubes. Upon rotation of the spiral (a helical coil), crystals form in a freezing section at the top of the column and are forced downward against the warmer liquid section at the bottom of the tube. The countercurrent movement of crystals and melt comprises a multistage fractional crystallizer. The differences in solid and liquid composition in this continuous crystallization resemble the separation of liquid and vapor phases in fractional distillation.

After the addition of a typical feed (benzene, e.g.) and the attainment of equilibrium, the spiral drives crystalline benzene down as liquid rises between the tubes. High purity product is drawn off from the melting section at the bottom. The high-purity liquid washes impurities from the surface of the descending crystals. The smoothness of the spiral surface is important to promote effective crystal transport.[88] Desirable ratios of spiral pitch to annulus width to column diameter are 3/1/8. Continuous column purification on a micro scale can be carried out when only small quantities are available for ultrapurification.[89] Atwood recently presented a comprehensive review of this technique.[90]

D. CRYSTALLIZATION

Crystallization is the most common procedure for the purification of solids, particularly for the removal of insoluble particles.[91, 92] A saturated, boiling solution of a solid in an appropriate solvent is filtered through a 0.8 μm membrane filter under reduced pressure. The filtrate is allowed to cool to room temperature or below, permitting the dissolved substance to crystallize out. Ideally, the substance to be purified should be much more soluble at higher temperatures than at room temperature.

For example, 100 g of potassium dichromate can be dissolved in 100 ml of water at 100°C. After filtration and cooling, the bulk of the solute crystallizes. At 0°C only 4.9 g is soluble in 100 ml. The purified product should be filtered off on a polyethylene funnel and dried at 150°C *in vacuo* to remove water. Some solids such as sodium chloride possess roughly the same solubility from 0° to 100°C. The solubility of NaCl in water is 35.7 and 39.1 g/100 ml at 0 and 100°C, respectively. In this case a near-saturated solution is filtered; the filtrate is concentrated to one-half or one-third the original volume *in vacuo* and cooled before the purified product is collected. The crystals of sodium chloride occlude water during their formation. When this salt is to be used as a primary analytical standard, it must be dried at 600–650°C for complete removal of water. Conventional drying at about 120°C is unsatisfactory.

E. EXTRACTION AND COMPLEXATION

Organic reagents that form chelate complexes are important in many aspects of analytical chemistry. Excellent reviews are available on the use of chelating agents as analytical precipitants, extractants, masking agents, and metal solubilizers.[93-97] These reagents can also be used for the effective removal of trace metals from a variety of matrices. Purification of the most important organic chelating agents that have found wide application in analytical chemistry is briefly summarized below.

1. Dithiocarbamates

Sodium diethyldithiocarbamate, readily water soluble (35 g/100 ml), is generally used as a 2% aqueous solution. Since its first application in the analysis of copper and iron in 1908,[98] this nonselective precipitant or extractant in the sodium form has been adopted for complete extraction of transition elements.[96] The pH of the aqueous solution is important because the sodium derivative of diethyldithiocarbamic acid decomposes in strong acid solution.[99]

The diethylammonium salt is preferred in the extraction of metals from highly acid media.[100] Reagent ammonium or sodium salts of dithiocarbamates can be purified by extracting 1% aqueous solution of the salt with a ketonic solvent (commonly methyl isobutyl ketone).[101] Luke has introduced the "Coprex" process, whereby trace metals are precipitated as the dithiocarbamate chelates with sodium diethyldithiocarbamate, collected on membrane filter disks, and determined by X-ray fluorescence measurements.[102, 103]

In recent years ammonium 1-pyrrolidinecarbodithioate has become one of the popular extractants. Malissa and Schoffman[104] showed that this reagent forms neutral, water-insoluble, solvent-extractable chelates with most polyvalent metals. For separation and concentration in atomic absorption spectrometry,[105, 106] this chelating agent must be low in multivalent metal content. It is easily synthesized from pyrrolidine, carbon disulfide, and ammonia. Carbon disulfide and pyrrolidine are purified by careful fractional distillation. Isopiestically distilled aqueous ammonia is the preferred base because of the purity attainable. Sodium and potassium hydroxide are ultrapurified with considerable difficulty, but ultrapure sodium carbonate is applicable for the sodium salt.

The free acid, prepared by dissolving purified pyrrolidine and carbon disulfide in chloroform, has a much greater solubility in this solvent than any corresponding salt in water. First 50 ml of chloroform is added to a 100 ml volumetric flask. Pyrrolidine (4.3 ml) is then pipetted into the flask, which is placed in a hood. While the flask is swirled, carbon disulfide (3.0 ml) is added in drops. The solution is cooled to room temperature and diluted to volume with solvent. This stock solution is 0.5 M in the reagent.

2. 8-Quinolinol

8-Quinolinol is a major extractant and precipitant that forms five-membered rings with metals. Although the reagent grade product is adequate for most analytical purposes, the blanks encountered in ultratrace work can become intolerable. 8-Quinolinol can be recrystallized from ethanol and water, then sublimed repeatedly to remove trace metals (see Sublimation, Section I.I). An ultrapure product prepared in this manner is now offered with a total polyvalent metal content of less than 1 $\mu g/g$.[52] When this product is incorporated in the 8-quinolinol–thionalide–tannic acid mixture with indium as the collector ion, improved spectrographic limits are attained.[107] Since the 8-quinolinol is the major part of this mixture on a weight basis, the absence of polyvalent metals in this reagent is essential.

Eckschlager[108] showed that at 410–600 nm the absorbance of 8-quinolinol was substantially reduced by zone refining and sublimation. We have found that many chelates of 8-quinolinol show strong absorbance in this area and that reduction in absorbance indeed parallels a reduction in the content of multivalent elements.

3. Dithizone

Diphenylthiocarbazone

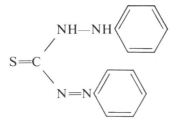

known as dithizone, was first applied to the analysis of metals by Fischer.[109] Several surveys of the conditions for isolation and determination of metals with this important reagent have appeared.[45, 96, 110, 111] The purplish-black crystalline powder decomposes at 167–169°C; it is insoluble in water and dilute mineral acids, but soluble in chloroform and carbon tetrachloride.

Since the commercial product is sensitive to oxidation, removal of the oxidation product, diphenylcarbadiazone, is usually required. Purification can be accomplished by filtering a saturated chloroform solution, then evaporating the solvent in a stream of filtered nitrogen at 40°C under clean-room conditions until one-half the original dithizone crystallizes out. The crystallized product is washed several times with carbon tetrachloride on a polyethylene filter, then dried *in vacuo*. Almost 100% pure product can be prepared by adding 3–4 volumes of petroleum ether to a filtered chloroform solution of dithizone and allowing the product to crystallize.[110] Alternatively, a chloroform solution of the reagent is extracted several times with isopiestically distilled aqueous ammonia ($\sim$ 0.2 M). Dithizone dissolves in the aqueous solution; the oxidation product remains in the organic layer. Acidification of the ammoniacal solution liberates the free, purified dithizone. Stock solutions containing 0.1 g of this purified product in 1 liter of purified chloroform are suitable for a variety of extraction procedures.

The purification procedures described previously are adequate for the removal of organic impurities. Reduction of trace metal content depends on the cleanliness of the Teflon separatory funnel during extraction, the purity of the chloroform, the nature of the interface between the water and chloroform layers, and the particle content in the laboratory atmosphere. Under favorable conditions the lead blank of the dithizone stock solution (0.1 g/liter) is 0.01 ng Pb/ml.[112] Solutions containing up to 1.7 g/ml are stable at 25°C.

This high purity stock solution is an excellent extractant for the purification of aqueous salt solutions functioning as buffers or electrolytes in analyses for trace elements reacting with dithizone (see Table 12). Reagent grade salts of cations nonreactive to dithizone are dissolved in high purity water and filtered through 0.2 μm cellulose acetate membrane filters in a Teflon (TFE) filter holder under nitrogen pressure (8–10 lb) to remove particulate matter. The filtrate is adjusted to pH 3–6 with high purity HCl or sodium carbonate and extracted two or three times in a Teflon (FEP) separatory funnel with dithizone stock solution. After several washes with high purity chloroform, the aqueous salt solution has unusually low blanks for the transition and heavy metals forming red, violet, or orange metal dithizonates soluble in the organic layer. Table 12 shows that aqueous solutions of alkali and alkaline earth salts, especially pure with respect to 23 elements, are theoretically possible in the laboratory. Chloride, nitrate, or acetate salts are preferable; cyanide is a metal-complexing agent that should be avoided. The correct pH for quantitative extraction of a specific element must be determined in each case.

4. Acetylacetone

Acetylacetone, 2,4-pentanedione, boils at 139°C (745 mm) and melts at −23°C. It can act as both solvent and extractant. Since the solubility of six-

TABLE 12. Elements Extracted by Organic Solutions of Dithizone[a]

H																	
Li	Be											B	C	N	O	F	
Na	Mg .											Al	Si	P	S	C	
K	Ca	So	Ti	V	Cr	Mn	Fe	Co	Ni	Cu	Zn	Ga	Ge	As	Se	Bi	
Rb	Sr	Y	Zr	Nb	Mo	Te	Ru	Rh	Pd	Ag	Cd	In	Sn	Sb	Te	I	
Ca	Ba	La	Hf	Ta	W	Re	Os	Ir	Pt	Au	Hg	Tl	Pb	Bi	Po	—	
—	Ra	Ac	Th	Pa	U												

[a] Elements enclosed within the solid lines are extracted.

membered metal acetylacetonates in organic solvents is considerable, macro scale separations are feasible.

The main impurity in commercial material is acetic acid. Extraction with (1:10) aqueous ammonia followed by deionized water provides purified material that is dried over anhydrous sodium sulfate, then distilled. An ultrapurified product is prepared by fractionally freezing 75–85% of the distillate.

5. PAN

1-(2-Pyridylazo)-2-naphthol (PAN) has been receiving increasing attention as a chromogenic and precipitating agent since its introduction in 1955.[113] The orange-red solid melts at 142°C, it is insoluble in water but soluble in strongly acid or alkaline aqueous solutions, as well as in a variety of organic solvents. It is prepared by diazotizing 2-aminopyridine with butyl nitrite and coupling with 2-naphthol in ethanol. Sublimation of the crude product and crystallization from ethanol produces a purified material.[114]

6. EDTA

(Ethylenedinitrilo)tetraacetic acid has become the most versatile chelating agent since World War II. It forms stable, water-soluble 1:1 chelates with a large number of polyvalent metal ions. The free acid is insoluble in water; thus the reagent is used as the disodium salt.

A variety of chemicals can be purified by precipitation or recrystallization in the presence of EDTA as a masking agent for trace metals. Lithium carbonate, for example, can be obtained virtually free of di- and polyvalent cations by adding free EDTA to a solution of lithium hydroxide and passing carbon dioxide through the solution.

Reagent grade EDTA is satisfactory for most laboratory applications, but high purity EDTA low in multivalent metals is needed as a masking agent in trace analysis and in the preparation of high purity chemicals. A complete procedure has been reported for the preparation of high purity EDTA as the free acid. The disodium (ethylenedinitrilo)tetraacetate solution is filtered through a 0.25 μm Millipore membrane filter and precipitated as the free acid from the filtrate by slow neutralization with high purity mineral acid.[115] Typical values for one lot of high-purity EDTA are summarized in Table 13. Sodium at 25 μg/g is the only metal impurity present above the 1 μg/g level, and some 28 polyvalent metals have been reduced to a total content of ≤ 2 μg/g. The high assay of this product permits its direct use as a chelometric standard. Complete experimental details are reported for this preparation and analysis of EDTA.[107]

TABLE 13. Certified Values for One Lot of High Purity EDTA[a]

Assay and Analysis		Metallic Impurities[b] (ppm)	
Assay ($C_{10}H_{15}N_2O_8$) after		Aluminum	0.03
drying at 105°C for 2 hr		Antimony	<1
Potentiometric acid-base	$100.0_3\%^c$	Barium	<0.04
weight titration		Beryllium	<0.01
Chelometric weight titra-	$100.0_4\%^d$	Bismuth	<0.1
tion with photometric		Cadmium	<0.2
indication[e]		Calcium	0.1
Elemental analysis		Chromium	<0.1
Carbon (theory 41.01%)	41.3%; 41.3%	Cobalt	<0.1
Hydrogen (theory 5.52%)	5.6%; 5.5%	Copper	0.02
Nitrogen (theory 9.59%)	9.6%; 9.6%	Gallium	<0.1
Ash (sulfated)	0.003%	Germanium	<0.2
Loss on drying at 105°C for	0.07%	Gold	<0.2
2 hr		Indium	<0.02
Nitrilotriacetate content[f]	0.02%	Iron	0.1
Particulate matter (after dis-	0.003%	Lead	<0.2
solution in aqueous		Magnesium	0.01
ammonia)		Manganese	<0.02
Nonmetallic impurities (ppm)		Molybdenum	0.02
Arsenic[g]	0.05	Nickel	<0.1
Boron[b]	0.02	Niobium	<0.2
Halide (as Cl)	30	Silver	0.01
Silicon[b]	0.01	Sodium	25
		Strontium	0.1
		Tin	<0.1
		Titanium	0.02
		Vanadium	<0.02
		Zinc	<2
		Zirconium	<0.1

[a] Actual lot analysis, lot UHC331, ULTREX® (ethylenedinitrilo)tetraacetate acid, J. T. Baker Chem. Co., as packaged under argon in vials. The data support assignment of a 100.0% EDTA value to the product as a chelometric standard.
[b] By d, c-arc spectrography.
[c] Average value for 5 determinations: 100.1_0, 99.9_2, 100.0_3, $100.0_3\%$, and $100.0_6\%$.
[d] Average value for 3 determinations: 100.0_6, 100.0_4, and $100.0_3\%$.
[e] Weighed amount of ULTREX® calcium carbonate, lot UMO 450, assay 99.99% $CaCO_2$: reacted with a substoichiometric weighed amount of the EDTA. The small excess of calcium in the solution, ammoniacally buffered to pH 9.5, titrated photometrically with a standard EDTA solution, with Calmagite as indicator.
[f] By dc-polarographic assessment of the cadmium–NTA wave (−0.94V vs. SCE).
[g] By evolution and silver diethyldithiocarbamate photometry.
Source: Ref. 115.

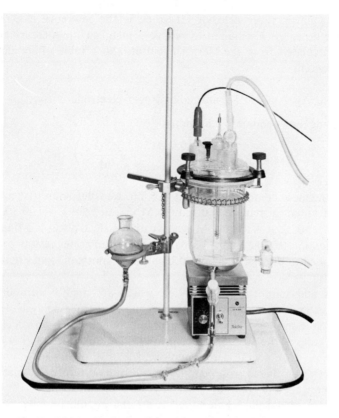

Fig. 7.   Mercury cathode cell for ultrapurification of electrolytes.

F. ELECTROLYSIS

Controlled potential electrolysis at a mercury cathode is an efficient, simple, and rapid method for the removal of trace metals from aqueous solutions.[116] Electrolytic cells for "polishing" electrolyte solutions have become available* (Figure 7). Here electrolysis is carried out with a silver/silver chloride anode and a mercury pool cathode under a blanket of inert gas (nitrogen or argon). The removal of manganese from 1 M solution of calcium nitrate by mercury cathode electrolysis demonstrates the utility of the method.[117] When 4000 ml of the calcium nitrate solution was electrolyzed with 234 ml (3.2 kg) of mercury functioning as cathode, 99% of the manganese was removed from the aqueous solution.

* Environmental Science Associates, Burlington, Mass.; Princeton Applied Research, Princeton, N.J.

When a silver/silver chloride electrode is the reference electrode, the minimum potential E required to reduce a metal at the mercury electrode can be calculated from the Nernst equation and a table of standard electrode potentials.

$$E_{vs} \text{ Ag/AgCl} = E_{vs} \text{ normal hydrogen electrode} - E_{Ag/AgCl} \qquad (1)$$

For manganese, equation 1 becomes

$$E_{Ag/AgCl} = -1.18 \left(\frac{0.05915}{2} \log \frac{1}{Mn^{2+}} \right) - 0.22 \qquad (2)$$

If the manganese content in 1 M calcium nitrate solution is 10^{-6} molar, the minimum potential required to reduce MN^2 to Mn $= -1.18$ $(0.05915/2$ $\log 1/10^{-6})-0.22 = -1.58$ V. Mercury cathode electrolysis is a final polishing purification that is retained for prepurified substrates just as zone melting is applied to ultrapurification of high purity inorganic and organic compounds.

Buffers and supporting electrolytes, which are readily ultrapurified by mercury cathode electrolysis, are often required in trace analysis. In anodic stripping voltammetric procedures it is common to analyze for lead, cadmium, or zinc at the 10^{-6} M concentration in a 1 or 2 M solution of a buffer or simple salt. In the determination of lead in blood, for example, the whole blood is refluxed with perchloric acid to oxidize organic matter, then 2 M sodium acetate solution is added before anodic stripping.[118] Since the lead content of normal whole blood is 0.3-0.4 μg/ml, roughly 1.5 $\times$ 10^{-6} M,[119] the amount of lead in the aliquot of reagent sodium acetate solution ideally should be 1.5 $\times$ 10^{-7} M, one-tenth of the value being determined. Solid sodium acetate, therefore, must contain less than 1 $\times$ 10^{-6}% lead, a value not attained in most commercially available reagents.

Sodium or potassium salts of acetic, citric, and boric acids, potassium acid phthalate, potassium dihydrogen phosphate, and dipotassium hydrogen phosphate are popular salts in buffer systems. In addition, potassium and sodium chloride are preferred supporting electrolytes that can be readily ultrapurified.

Electrolysis of dissolved sodium carbonate was carried out with a platinum anode and mercury cathode under an inert gas blanket (argon) at a potential of -1.5 V versus a standard calomel electrode. Concentrations of chromium, copper, iron, and nickel in the electrolyzate amounted to < 7, 10, 10, and 10 ng/g of the carbonate, respectively.[120] The purest sodium hydroxide (50% solution) has been prepared commercially by the electrolysis of an aqueous solution of sodium chloride in a cell containing a mercury cathode and a graphite anode. The sodium is reduced at the cathode and forms a dilute amalgam. The amalgam is then removed to a separate chamber, where it reacts with water to form sodium hydroxide solution. Traces of mercury are found in this mercury cell product. In addition, iron contamination results during the transport of the solution in stainless steel tanks. In concentrated alkali, iron exists as a ferrate ion that is difficult to remove in the average laboratory. However a sample* of 50% mercury cell sodium hydroxide that was further purified by passage through a flowthrough, continuous electrolytic cell based on a porous carbon cathode possessed only 20 ng/g of iron; mercury was not detectable.[121]

Although mercury cathode electrolysis provides an excellent tool for the isolation of ultrapure chemicals, unusually rigid handling procedures are required to preserve the integrity of the product. The electrolysis should be carried out in a laminar-flow hood. Wherever possible, polyethylene, polypropylene, or polyfluorohydrocarbon equipment should be chosen. After electrolysis, the purified solution is siphoned into a Teflon (TFE) container from which it is pressure filtered through a 0.2 μm cellulose acetate filter supported on a polyethylene frit.

G. FILTRATION, A PREPURIFICATION TOOL

1. Purification of Fluxes

The value of membrane filtration under pressure has been demonstrated in the preparation of alkaline fluxes for attacking refractory materials such as silicates, mineral oxides, and some iron alloys.

The purest alkaline flux is sodium carbonate. A 25% (w/v) solution of reagent grade material (Fe = 2 μg/g) was prepurified by filtration through a Whatman #42 paper to remove particulate matter and insoluble heavy metal salts. The iron content in the filtrate was reduced to 0.20 μg/g based on Na_2CO_3. When the filtrate was passed through a 0.2 μm cellulose acetate membrane under 30 psi. of argon pressure, the iron concentration was found to be 0.015 μg/g by the adaptation of ferrozine spectrophotometry[91] to ultratrace analysis.[43] It is evident that undesirable metal carbonates insoluble in alkaline solution can be removed effectively by selection of the proper filtration conditions.

In the F. Laurence Smith method for the determination of alkali metals in silicates, a mixture of ultrapure calcium carbonate and ammonium chloride is employed.[122] Upon heating, this mixture liberates CaO and $CaCl_2$, the reaction products that actually dissolve silicates. Calcium carbonate is prepared by the precipitation of calcium nitrate with ammonium carbonate. Ammonium chloride is prepared from isopiestically distilled ammonium hydroxide and hydrochloric acid.

* Supplied by Dr. G. A. Carlson, PPG Industries, New Martinsville, W. Va.

2. Purification of Calcium Nitrate

A 2 *M* solution of reagent grade calcium nitrate tetrahydrate was filtered through a Whatman #42 paper. The iron content was reduced from 5.0 to 0.3 $\mu g/g$. Pressure filtration through a 0.2 μm cellulose acetate filter, as described earlier for sodium carbonate, reduced the iron to 0.07 $\mu g/g$.

Although the size of particles in the filtrate can be correlated with mean pore size of the membrane, a small number of particles larger than the rating of the membrane do appear in the purified liquid. For example, some 5 μm particles pass through a 0.2 μm porosity polymeric membrane (see Table 10). Porous carbon tubes coated with an inorganic composite are being introduced for the separation of particles 0.001–0.005 μm in diameter.[123] If such separations can indeed be attained in laboratory preparative work, simple pressure filtration becomes a very valuable prepurification step for water-soluble salts, particularly those that dissolve to give a pH 5.0 or higher.

H. PRECIPITATION

Coprecipitation is a valuable prepurification method. The coprecipitation of trace impurities on "carrier" precipitates can be applied to at least 50 elements.[124, 125] Inorganic traces other than alkali or alkaline-earth cations can be coprecipitated as the oxide or hydroxide from sodium, potassium, calcium, magnesium, and barium salts, since these matrices form soluble hydroxides. Iron, aluminum, titanium, and lanthanum[126] have been used as carriers. Lanthanum is the carrier of choice because it is not normally under investigation in ultratrace analysis. Iron, on the other hand, is one of the important transition elements that must be controlled at the nanogram level. The following purification of calcium nitrate is typical of the method.

1. Prepurification of Calcium Nitrate

A 2 *M* solution of reagent grade calcium nitrate is filtered through Whatman #42 paper *in vacuo*. The iron is reduced from 3 to 0.3 $\mu g/g$.

2. Ultrapurification of Calcium Nitrate

Prepare a solution of 2.0 g reagent grade lanthanum nitrate hexahydrate in 100 ml, and add 10 ml of this solution to 500 ml of the prepurified 2 *M* calcium nitrate solution. After the addition of 4 drops of 30% hydrogen peroxide to oxidize iron and chromium, the pH is adjusted to 7.5 with 10% ammonium hydroxide and the solution is heated to 60°C. After cooling to room temperature, the lanthanum precipitate containing the coprecipitated trace hydroxides is filtered off. The iron content in the filtrate is reduced from 0.3 to 0.01 $\mu g/g$ based on the calcium nitrate concentration. The gelatinous lanthanum precipitate filters somewhat slowly.

Purification can also be carried out by adding 0.1 g of aluminum chloride hexahydrate to the same volume of calcium nitrate solution and following the identical procedure except that the pH is adjusted to 6.0 with ammonium hydroxide. Again the iron content of the filtrate becomes 0.01 $\mu g/g$ (analyzed by extraction with ammonium 1-pyrrolidinedithioate in MIBK, followed by atomic absorption spectrophotometry of the extract). Coprecipitation, electrochemical reduction, or chelating resins are equally effective in reducing the iron to this level.

Alkali and alkaline-earth salts can also be purified via sulfide precipitation. The addition of indium chloride to aqueous solutions of these elements and the subsequent addition of hydrogen sulfide coprecipitates trace sulfides with the indium sulfide carrier.

Coprecipitation of the hydroxide or sulfide is nonselective, generally applicable to group separations. Specific separations can also be devised. If lead is a trace contaminant, the addition of barium chloride and sulfuric acid to an aqueous salt solution coprecipitates lead sulfate with the barium sulfate carrier. Traces of nickel can be coprecipitated with copper dimethylgloxime after the addition of copper chloride and dimethylgloxime.[127]

I. SUBLIMATION

The apparatus, techniques, and applications for sublimation have been critically reviewed.[128, 129] The value of sublimation in the production of free-flowing, crystalline products from the solid chunks isolated after zone melting has already been mentioned (see Zone Melting, Section I.C.2). Although sublimation equipment in the laboratory is adequate only for small quantities, a novel low-cost procedure for the preparation of phosphorus pentoxide in quantity has been described.[130]

When low-lead phosphoric acid for the dissociation of Apollo 11 lunar samples was required,[131] the low volatility of this acid precluded the distillation techniques successful with HCl and HNO_3. The problem was resolved by purification of phosphorus pentoxide and subsequent mixing of this oxide with high purity water. Thrice-sublimed phosphorus pentoxide had a lead content of less than 0.1 ng/g by mass spectrometric analysis compared to 1–10 μg in typical reagent grade material.[132] Since the maximum limit for lead in ACS phosphorus pentoxide is 10^5 ng/g, sublimation has effected a millionfold reduction in this specification. The pentoxide is stable and can be stored in a sealed, precleaned borosilicate ampoule.

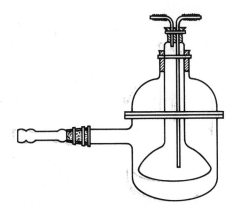

Fig. 8.   Laboratory sublimation apparatus.

8-Quinolinol that is routinely employed for trace collection in emission spectrographic analysis is obtained via sublimation. The following purification is illustrative of small-scale laboratory preparations applicable to a variety of inorganic and organic chemicals.

In the prepurification of 8-quinolinol, reagent-grade material (454 g) is dissolved in 900 ml of 95% hot ethanol containing 2 drops of concentrated HCl. The hot solution is decolorized by boiling with 9 g of acid-washed decolorizing carbon (Darco G-60)* for 10 min; then it is filtered and allowed to cool to room temperature. The crystals are filtered off and dried at room temperature *in vacuo*; the melting point is 75°C, and the yield is 320 g (71%).

To ultrapurify 8-quinolinol, place 30 g of prepurified material in the bottom of a sublimation apparatus† (Figure 8). A Teflon gasket sandwiched between the upper and lower sections is a good replacement for the grease normally applied. A silicone stopper is satisfactory for keeping the condenser in place. The bottom section is placed in a heating mantle controlled by a variable resistance. The mantle is mounted on a laboratory desktop covered with a polyethylene sheet. A "tent" of additional polyethylene sheeting encloses the apparatus except for the vacuum pump and manometer; the sheet prevents any dustfall from contaminating the product. After the apparatus is put together and cold water is circulated through the condenser, the system is evacuated to 1 mm by an oil pump. The sublimand is heated to 70°C, and the sublimate is collected on the underside of the condenser positioned about 5.0 cm above the top position of the prepurified material. The sublimation is continued until the underside of the condenser is covered with crystalline, colorless product. The vacuum is broken by

* Darco G-60, Atlas Chemical Co., Wilmington, Del.
† Scientific Glass Apparatus Co., Bloomfield, N.J.

introducing laboratory air filtered through a 47 mm–0.2 μm cellulose acetate, in-line membrane filter. The product is scraped off the condenser with a Teflon spatula into a precleaned borosilicate glass bottle provided with a Teflon-lined cap; yield is 28.5 g (95%).

II. GASES

Highly pure gases, especially helium, hydrogen, argon, nitrogen, air (with respect to organics), ammonia, acid vapors, and volatile compounds, are frequently required in the preparation of reagents for trace analysis or in the production of various semiconductor devices and optical wave guides by chemical vapor deposition (CVD). In these applications gases or vapors must be purified to remove traces of other gaseous impurities. It is also necessary to prevent particulate contamination and absorption of moisture.

High purity gases are primarily contaminated during storage, transportation, and metering. It cannot be overemphasized that careful handling of a high purity gas is the key to maintaining quality. The successful use of high purity gases in practical work without contamination requires stringent controls. Since the system is immersed in the contaminant (air, water vapor, etc.), working with high purity gases has been compared to performing a trace water determination under water.

This section reports the problems and pitfalls occurring during the industrial production and laboratory purification of high purity gases. Information on proper handling techniques is also provided.

A.   INDUSTRIAL PRODUCTION AND PURIFICATION*

Industrial gases produced domestically in billion cubic foot quantities per year (nitrogen, 300 billion; argon, 3.3 billion; and hydrogen, 450 billion ft^3) contain only trace amounts of other gaseous contaminants.[133] Typical gaseous impurities in industrial gases (as produced) are shown in Table 14. In addition to trace gaseous impurities, the gas from the manufacturing unit is always contaminated to some degree with water (100–1000 ppm), air, particulate matter (rust), and sometimes with oil and other organic material that is introduced when the gas is compressed in the cylinder (see Table 15).

The impurity levels of the as-produced gas are increased during handling. By the time of use, about 90–99% of the total contamination in the gas has been introduced by the manufacturer and the user. The prime source of contamination by the manufacturer is storage in the compressed gas

* The authors are indebted to C. A. McMenamy, J. T. Baker Chemical Co., for the extensive information provided on this topic.

TABLE 14.   Principal Impurities in Industrial Gases

Gas and Purity[a]	Impurities
Argon, 4N8	Nitrogen 2 ppm, oxygen 0.2 ppm, methane 1 ppm
Helium, 5N	Neon 15 ppm, nitrogen 5 ppm, oxygen 0.5 ppm
Hydrogen, 4N5[b]	Helium 10 ppm, nitrogen 2 ppm, oxygen 0.1 ppm
Nitrogen, 4N	Argon 100 ppm, neon 3 ppm, oxygen 1 ppm
Oxygen, 4N5[c]	Methane 20 ppm, krypton 15 ppm, nitrogen 10 ppm, argon 20 ppm

[a] 5N = 99.999%; 4N5 = 99.995%.
[b] Product from natural gas re-forming–liquid hydrogen plant.
[c] Commercial oxygen (2N7) contains lower krypton and methane. The argon content is much higher (2000–3000 ppm).

cylinder.[134] Innovations in the design of rustproof containers for high purity gases have been impeded by economic considerations. However one supplier of industrial gases intends to introduce anodized aluminum cylinders.*

Impervious polyfluorocarbon-coated cylinders should be developed for highly corrosive gases. Although storage is a primary concern in preventing contamination, other general considerations are also applicable to the proper commercial handling of high purity gases. Several general rules have been provided by C. A. McMenamy.[135]

1.   Clean metal tubing and metal diaphragm regulators should be used.

2.   Sintered metal filters or a Millipore filter system should be inserted in the line to reduce particulate matter.

TABLE 15.   Contaminants Introduced During Handling

Source	Contaminant	Cause
Cylinder	Water	Improper cylinder drying
	Rust	Corrosion of the inner cylinder wall
	Oil	Oil lubricated compressor used for filling
	Other	Contaminants introduced in cylinder from a previous user
Cylinder valve	Air	Leaks
	Organics	Thread compound
Regulator	Air	Leaks, diffusion
	Water	Inadequate purge
	Oil	Contaminated with oil

* AirCo, Murray Hill, N.J.

TABLE 16.   Purity Ranges of Industrial Gases

Gas	Purity as Produced	Number of ·Grades	Lowest Purity	Highest Purity
Argon	4N8	8	3N8	5N
Helium	5N	7	2N	6N
Hydrogen	Variable	10	1N5	5N7
Nitrogen	4N	14	2N	5N
Oxygen	2N7	—	2N	5N

Source: Compressed Gas Association.

3.   Fittings and tubing must be washed with a degreasing solvent such as trichloroethylene.

4.   Teflon tape only should be applied to pipe fittings.

5.   Compressed gases should always be tested individually (i.e., check each cylinder) for water content before use. Liquid water in the cylinder causes the water level to increase as the cylinder pressure decreases. When a moisture-free system is required, stainless steel is preferable to copper tubing.

6.   A gas handling system should be purged with dry nitrogen to remove water before use. Heating also helps during the purge.

A variety of different grades of gases are available commercially. Some of the designated purity ranges of several industrial gases appear in Table 16. These specifications should only be related to the as-produced gas. In most cases the *difference between commercial grades is a matter of care in handling* and possibly the degree of characterization by analysis.

Economic methods for the industrial purification of gases are based primarily on the use of selective or general purpose absorbants* to remove unwanted impurities. The majority of trace gaseous impurities in inert

* Three types of molecular sieves (crystalline sodium or calcium aluminosilicates) are offered by the Linde Division of Union Carbide Corp.

Type	Pore Diameter	Molecules Adsorbed
4A	4A	Water plus other small molecules (CO_2, H_2S, SO_2, NH_3, C_2H_6); below -30 C., N_2, O_2, and CH_4.
5A	5A	Small molecules listed above *plus* larger molecules such as butane, hexane, *n*-butanol, and higher *n*-alcohols. *Generally used for drying gases* and low temperature adsorption of impurities.
13X	10A	Previous substances listed *plus* branched chain and cyclic materials.

TABLE 17. Low Temperature Purification of Gases with 5A Molecular Sieves

Gas	Coolant	Principal Impurities Removed
Hydrogen[a]	Liquid nitrogen	Nitrogen, argon
Helium	Liquid nitrogen	Nitrogen, argon, oxygen
Nitrogen	Liquid oxygen	Carbon monoxide, oxygen
Argon	Liquid oxygen	Nitrogen, methane
Oxygen	Solid carbon dioxide/Solvent	Ethane, carbon dioxide, krypton

[a] Hydrogen should be passed through a Deoxo trap before the cryogenic trap to prevent oxygen buildup.

industrial gases can be adsorbed on 5A molecular sieves at cryogenic temperatures. The principal impurities removed with different coolants by a trap loaded with this absorbant are listed in Table 17. When a trap of 1.2 cm i.d. × 4.5 cm length is cooled in liquid nitrogen to −196°C, about 300 liters of hydrogen can be purified. The original nitrogen content (5 ppm) decreases to less than 0.1 ppm. Trace impurities can be concentrated by this technique in a cryogenic adsorption trap and measured with a conventional gas chromatograph (see Table 18).

Methods based on chemical reaction of contaminants with metals or adsorption media are economic for industrial purification. Several techniques utilize the chemical decomposition of impurity and subsequent absorption of the reaction products. The gas stream is passed at high temperature over a suitable medium, and decomposition products are trapped cryogenically or chemically as previously described. Some typical impurity levels before and after purification of inert gases by these methods are given in Table 19. The level of impurities in the gas entering the reaction chamber or trap (1.2 cm i.d. × 15 cm long) at a flow rate of 500 cm³/hr is compared with the impurity level of the gas emerging from the trap.

TABLE 18. Lowest Measurable Concentration (ppm) of Traces[a]

Impurity	Volume of Gas Sample Passed Through Concentrator			
	10 cm³	100 cm³	500 cm³	2000 cm³
Argon	5	0.5	0.1	0.03
Nitrogen	10	1	0.2	0.06
Carbon dioxide	20	2	0.4	0.02

[a] In hydrogen-UHP hydrogen carrier gas; Beckman GC 2A Gas Chromatograph equipped with a thermal conductivity detector (300 M.A.) and a cryogenic concentrator filled with 5A molecular sieves.

TABLE 19. Removal of Trace Components from High Purity Gases

Impurity	Amount (ppm)		Method of Purification
	Before	After	
Hydrogen	5	0.2	Absorption on palladium
Oxygen	2	0.1	MnO catalyst
Carbon dioxide	2	0.1	Molecular sieves, 5A
Ethane	0.5	0.01	Combustion over CuO/molecular sieve
Water	2	0.1	Molecular sieves, 5A, highly activated

Hydrogen can be removed by a "Deoxo"-type palladium on alumina catalyst or with palladium loaded molecular sieves. Oxygen is adsorbed on manganese oxide formed as a coating on alumina by oxidation of manganous nitrate followed by hydrogen reduction to the monoxide. Ethane is oxidized over cupric oxide wire at 900°C, and the reaction products (carbon dioxide and water) are adsorbed on high activity 5A molecular sieves. Type 5A molecular sieves also adsorb carbon dioxide, ammonia, and hydrogen sulfide from gas streams at room temperature.

B. LABORATORY METHODS OF PURIFICATION

Purification of gases and vapors before their use in trace analysis or in technologies requiring ultrapurity techniques is a routine requirement. Although the total quantity of gases used in these applications can be quite small, the success of the analysis or the results of the experiment are frequently critically dependent on the quality of the selected gas and the cleanliness of the storage and delivery systems.

As evident from the information of the industrial production and purification of gases, some minimum precautions must be exercised by the analyst or researcher to ensure that his purity specifications are being met for a particular application at the time of use of the purchased gas. The obvious general precautions are as follows:

1. Micron-sized filters to remove particulates.
2. A suitable unit for oil retention.
3. Adequate methods for removal of moisture.
4. Specific analysis or tests to detect impurities.
5. Purification methods for removal of impurities.

Methods for the selective absorption of gases before their determination by chemical or instrumental means can also serve to remove these trace gaseous impurities from other gases. Some common absorption media and the gases for which they are selective are listed in Table 20.

Oxygen reacts quantitatively with alkaline pyrogallol solution, alkaline sodium dithionite, or Cr (II) Cl_2 solution. Pyrogallol (1,2,3-trihydroxybenzene) is effective in alkaline solution at temperatures greater than 18°C. A mixture of 1:3 (15% aqueous pyrogallol:30% KOH) is used. The chromium chloride solution absorbs oxygen extremely rapidly but is more difficult to prepare and regenerate. Oxygen removal from alkaline solutions is accomplished by the addition of sulfite and hydrazine. The former can also be used in many neutral unbuffered solutions.

Dissolved oxygen must usually be removed from solutions that will be investigated by polarography. Most often solutions are deaerated by passing oxygen-free gas through the solution before polarograms are recorded; an air-free atmosphere is maintained over the surface of the solution during experimentation. Hydrogen, nitrogen, argon, and CO_2 (acidic solutions) containing less than 20 ppm of O_2 are employed.

Oxygen is effectively removed from a gas stream by passage through a tube filled with sulfur-free copper wire heated to 450–500°C. During purification of hydrogen, trace oxygen is converted to water vapor. In the case of nitrogen and other gases, CuO is formed. Other suitable purification methods are available. For example, oxygen is removed from hydrogen by passing the gas through finely divided palladium catalyst at room temperature. Vanadous solutions for oxygen absorption can be prepared by boiling 2 g of ammonium metavanadate with 25 ml of concentrated hydrochloric acid, diluting to 200 ml, and shaking with a few grams of amalgamated zinc.[136] Chromous solutions are prepared from a 0.5F solution of $CrCl_3$ in concentrated HCl. Amalgamated zinc is used for reduction of the chromic ion.

TABLE 20.   Selective Absorption of Gases

Gas	Absorption Media
Oxygen	1,2,3-trihydroxybenzene (Pyrogallol) (15% Pyrogallol-30% KOH, 1:3) or $CrCl_2$ solution
Hydrogen	Palladium catalyst, CuO at 280–300°C
Carbon dioxide	Potassium hydroxide solution (30–40%) Barium hydroxide
Carbon monoxide	Ag_2SO_4 in H_2SO_4, CuCl (ammonia)
HCl, NO_2, H_2S, SO_2 acetylene	Strongly alkaline solutions

Up to 1000 ppm of hydrogen may be removed as water from gases by passage over CuO heated in a quartz pipe at 225–280°C. Dissolved or dispersed oxidation agents such as silver permanganate[137] and sodium chlorate also oxidize hydrogen to water.

Technical grade hydrogen usually contains oxygen, nitrogen, and methane. These impurities can be removed by appropriate reactions discussed in this section. A method for the removal of $5 \times 10^{-5}\%$ oxygen in a 1.3 liter sample of hydrogen has been described.[138]

Carbon dioxide reacts completely with alkali and alkaline earth metal hydroxides in aqueous solution.[137] Potassium hydroxide is preferable to sodium, since the reaction is faster and potassium carbonate is more soluble than sodium carbonate. Strongly alkaline solutions also retain acidic or hydrolyzable gas components (HCl, NO_2, H_2S, SO_2, and acetylene).

Carbon monoxide is easily absorbed by ammoniacal and hydrochloric acid solutions of CuCl. Up to 1000 ppm of hydrocarbons may also be remove by reduction with hydrogen over nickel catalysts to produce methane.

Sulfur-containing gases, particularly SO_2, CS_2, and COS, can be oxidized with alkaline solutions of bromine or hydrogen peroxide to give sulfuric acid. Hydrogen sulfide (down to 1 ppb) can be adsorbed on suspended cadmium hydroxide.[139] Removal of SO_3 from SO_2 has been accomplished by selective absorption in a 4:1 mixture of isopropanol and water.[140]

Acetylene can be removed by reaction with several heavy metals to form stable acetylides. In Cu(I) solutions, a dark red precipitate of copper acetylide forms. Mercury (I) nitrate reacts with acetylene in ammoniacal N-methyl pyrrolidone solution.

Purification of helium is based on the diffusion of the gas through porous quartz. A quartz diffusion cell housed in a high pressure stainless steel tube has been used as the primary unit for the laboratory production of high purity helium. The tube is located in an electrically heated furnace with a built-in temperature controller for maintaining constant temperature regardless of flow conditions. A commercial unit* is reported to be useful for continuously producing helium with as low as 0.5 ppm of impurities. No periodic regeneration or liquid nitrogen-cooled adsorbants are necessary. These purifiers are particularly useful in gas chromatography.

Economical and convenient sources of pure air are used in gas chromatography to support combustion for flame ionization detectors (FID) and as a carrier gas.[141, 142] This application calls for air free from organic materials. Air can be purified batchwise with a modified heatless dryer.[143] When carbon is used as a purification medium, methane that occurs

* Matheson Gas Products, East Rutherford, N.J.

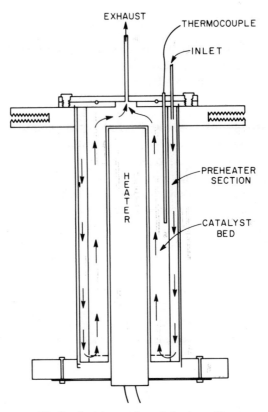

Fig. 9. Stainless steel catalytic air purifier.

TABLE 21. Drying Agents for Gases

Drying Agent	Water Content (g/m³) After Drying at 25°C
P_2O_5	0.00002
$MgClO_4$	0.0005
KOH (molten)	0.002
H_2SO_4 (100%)	0.003
Alumina	0.003
Molecular sieves	0.01–0.0001[a]
NaOH (molten)	0.16
CaO	0.2
$CaCl_2$	0.14–0.25[a]
H_2SO_4 (95%)	0.3
Silica gel	0.5[a]

[a] Varying values are given in the literature. The most effective methods depend on the use of granular phosphorus pentoxide or on a very dry molecular sieve trap (Type 4A activated for 12 hr at 350°C with a dry gas purge[147]).

144

naturally in air at concentrations up to 10 ppm by volume is not eliminated. Other work demonstrates that catalytic oxidation is ineffective for removal of chlorinated hydrocarbons from air.[144, 145] A successful catalytic oxidation method that delivers a constant flow of air free from organics has been described.[146] The catalytic purifier in Figure 9 has been used with 0.5% Pd on alumina. By including an oil aerosol filter* ahead of the catalytic burner, such systems can be used for 2 years without changing the catalytic bed.

Traces of water are routinely removed from various gases by passage through drying agents placed in a number of U-pipes. Table 21 presents the most common agents and their approximate effectiveness for water removal.[148]

REFERENCES

1. M. Zief, *Ind. Res.,* **5,** 36 (1971).

2. M. Zief, *Am. Lab.,* **10,** 55 (1971).

3. H. G. Griffin and T. D. George, Paper 72 presented at the Pittsburgh Conference, Cleveland, March 1972.

4. J. W. Mitchell, *Anal. Chem.,* **45** (6), 492A (1973).

5. ULTREX® Products, Catalog 750, J. T. Baker Chemical Co., Phillipsburg, N.J., 1970.

6. W. Wilcox, R. Friedenberg, and N. Back, *Chem. Rev.,* **64,** 194 (1964).

7. P. Jannke, J. K. Kennedy, and G. H. Moates, "Ultrapurification," in M. Zief and W. R. Wilcox, eds., *Fractional Solidification,* Dekker, New York, 1967, pp. 463–473.

8. E. H. Archibald, *The Preparation of Pure Inorganic Substances,* Wiley, New York, 1932.

9. D. D. Perrin, W. L. F. Armarego, and D. R. Perrin, *Purification of Laboratory Chemicals,* Pergamon Press, Oxford, 1966.

10. I. M. Kolthoff, E. B. Sandell, E. J. Meehan, and S. Bruckenstein, *Quantitative Chemical Analysis,* 4th ed., Macmillan London, 1969, pp. 435–451.

11. L. Vanino, *Handbuch der Preparativenchemie,* B and I, Anorganische Teil, Enke, Stuttgart, 1921.

12. M. Zief and R. Speights, eds., *Ultrapurity: Methods and Techniques,* Dekker, New York, 1972, Appendix.

13. G. Brauer, ed., *Handbook of Preparative Inorganic Chemistry,* Vol. 1, 1963; Vol. 2, 1965, Academic Press, New York.

14. A. F. Armington, M. S. Brooks, and B. Rubin, *Purification of Materials,* *Ann. N.Y. Acad. Sci.,* **137,** Art. 1, 1-402, January 20, 1966.

15. B. D. Stepin, I. G. Gorshteyn, G. Z. Blyum, G. M. Kurdycmov, and I. P. Ogloblina, *Methods of Producing Superpure Inorganic Substances,* Len-

---

* Deltech Engineering Inc., New Castle, Del.

ingrad, 1969. (Reproduced by National Technical Information Service, Springfield, Va., JPRS 53256.)

16. R. L. Globus and G. V. Chuchkin, *Zavod. Lab.*, **26**, 395 (1960).

17. M. A. Leipold and T. H. Nielson, *Ceram. Bull.*, **45** (3), 281 (1966).

18. A. K. De and S. H. Khopkar, *J. Sci. Ind. Res. (India)*, **21A** (3), 131 (1962).

19. F. P. Treadwell and W. T. Hall, *Analytical Chemistry*, Vol. I, 9th ed., Wiley, New York, 1937, p. 418.

20. *Tech. News Bull., Nat. Bur. Stand.*, **56** (5); 104 (1972).

21. E. C. Kuehner, R. Alverez, P. J. Paulsen, and T. J. Murphy, *Anal. Chem.*, **44**, 2050 (1972).

22. K. D. Burrhus and S. R. Hart, *Anal. Chem.*, **44** (2), 432 (1972).

23. J. M. Mattinson, *Anal. Chem.*, **44** (9), 1716 (1972).

24. P. C. Coppola and R. C. Hughes, *Anal. Chem.*, **24**, 768 (1952).

25. K. Little and J. D. Brooks, *Anal. Chem.*, **46** (9), 1343 (1974).

26. H. Irving and J. J. Cox, *Analyst*, **83**, 526 (1958).

27. W. Kwestroo and J. Viser, *Analyst*, **90**, 297 (1965).

28. C. L. Luke, Bell Telephone Laboratories, personal communication, 1973.

29. M. Tatsumoto, *Anal. Chem.*, **41**, 2088 (1969).

30. M. Zief and J. Horvath, unpublished results.

31. GC-Spectrophotometric Solvents, Catalog 750, J. T. Baker Chemical Co., Phillipsburg, N.J., 1975.

32. P. W. Byrnes, *Ind. Res.*, **15** (6), 40 (1973).

33. M. Winstead, *Reagent Grade Water: How, When and Why?* The Steck Co., Austin, Texas, 1967.

34. J. V. Jordan and G. D. Wyer, *Chem. Anal.*, **48**, 39 (1959).

35. *Duolite Ion-Exchange Manual*, Diamond Shamrock Chemical Co., Resinous Products Division, P.O. Box 829, Redwood City, Calif., 1969, p. 204.

36. R. M. Wheaton and A. H. Seamster, in *Encyclopedia of Chemical Technology*, Wiley, Vol. 11, 1966, p. 881.

37. R. Kunin, *Ion-Exchange Resins*, Wiley, New York, 1958.

38. D. M. Stromquist and A. C. Reents, *Ind. Eng. Chem.*, **43**, 1065 (1951).

39. J. H. Payne, H. P. Kortschak, and R. F. Gill, Jr., *Ind. Eng. Chem.*, **44**, 1411 (1952).

40. E. Blasius and B. Brozio, "Chelating Ion-Exchange Resins," in H. A. Flaschka and A. J. Barnard, Jr., eds., *Chelates in Analytical Chemistry*, Dekker, New York, 1967, pp. 49–79.

41. L. R. Morris, R. A. Mock, C. A. Marshall, and J. H. Howe, *J. Am. Chem. Soc.*, **81**, 377 (1959).

42. Duolite CS-346, Duolite Technical Sheet, Diamond Shamrock Chemical Co., Redwood City, Calif., June 1972.

43. L. L. Stookey, *Anal. Chem.*, **42** (7), 779 (1970).

44. Duolite Technical Sheet 114, Diamond Shamrock Chemical Co., Redwood City, Calif., May 1972.

45. *Amberlite Ion-Exchange Resins, Laboratory Guide,* Rohm and Haas, Philadelphia, 1971.

46. J. W. Mitchell, unpublished results.

47. S. A. Fisher and V. C. Smith, *Mat. Res. Stand.,* **12** (11), 27 (1972).

48. E. L. Gibbs, Ultrascience Inc., Evanston, Ill., 1972.

49. E. L. Gibbs, *In Vitro,* **8** (1), 37 (1972).

50. R. C. Hughes, P. C. Murau, and G. Gunderson, *Anal. Chem.,* **43** (6), 691 (1971).

51. S. A. Nazarey, A. J. Barnard, Jr., and J. A. Hesek, *Bull. Parenter. Drug Assoc.,* **28** (2), 70 (1974).

52. M. Zief and A. J. Barnard, Jr., *Chem. Technol.,* **3** (7), 440 (1973).

53. G. Bowers, Jr., Hartford Hospital, Hartford, Conn., personal communication.

54. J. W. Mitchell and D. L. Malm, unpublished results.

55. V. C. Smith, Ref. 12, p. 173.

56. K. A. Kraus and F. Nelson, Geneva International Conference on Peaceful Uses of Atomic Energy, Vol. 7, Session 9B, 1, 1955, p. 837.

57. W. J. Potts, Jr., *J. Chem. Phys.,* **20**, 809 (1952).

58. G. Hesse and H. Schildknecht, *Angew. Chem.,* **67**, 737 (1955).

59. G. Hesse, B. P. Engelbrecht, H. Engelhardt, and S. Nitsch, *Z. Anal. Chem.,* **241**, 91 (1968).

60. H. Brockmann and H. Schodder, *Ber.,* **74**, 73 (1941).

61. G. Hesse, *Z. Anal. Chem.,* **211**, 5 (1965).

62. G. Zweig and J. Sherma, eds., *Handbook of Chromatography,* Vol. II, CRC Press, Cleveland, 1972, p. 25.

63. F. M. Rabel, *Am. Lab.,* **7** (5) 53 (1975).

64. P. Pei, S. Ramachandran, and R. S. Henly, *Am. Lab.,* **7** (1), 37 (1975).

65. A. T. James and A. J. P. Martin, *Biochem. J.,* **50**, 679 (1952).

66. J. J. Kirkland, in *Gas Cromatography,* J. Coates, H. J. Noebels, and I. S. Fagerson, eds., Academic Press, New York, 1958, p. 203.

67. J. R. Gruden and M. Zief, Ref. 12, p. 131.

68. S. T. Sie, J. P. A. Bleumer, and G. W. A. Rijnders, *Separ. Sci.,* **1** (1), 41, (1966).

69. F. W. Schwab and E. Wichers, *J. Res. Nat. Bur. Stand.,* **32**, 253 (1944).

70. T. H. Gouw, *Separ. Sci.,* **3**, 313 (1968).

71. W. G. Pfann, *Zone Melting,* Wiley, New York, 1966.

72. E. A. Wynne, in *Fractional Solidification*, M. Zief and W. R. Wilcox, eds., Dekker, New York, 1967.

73. R. Albertins, W. C. Gates, and J. E. Powers, in *Fractional Solidification*, M. Zief and W. R. Wilcox, eds., Dekker, New York, 1967.

74. J. D. Henry, Jr., M. D. Danyi, and J. E. Powers, in M. Zief, ed., *Purification of Organic and Inorganic Compounds: Fractional Solidification*, Dekker, New York, 1969.

75. J. R. Gruden and M. Zief, Ref. 12, p. 125.

76. G. J. Sloan and N. Y. McGowan, *Rev. Sci. Instrum.*, **34** (1), 60 (1963).

77. R. W. Warren, *Rev. Sci. Instrum.*, **33**, 1378 (1962).

78. B. F. Oliver, *Trans. AIME*, **227** (4), 960 (1963).

79. A. Yamamoto and J. Akiyama, *Jap. Anal.*, **13**, 397 (1964).

80. E. A. Wynne, *Microchem. J.*, **5**, 175 (1961).

81. J. H. Beyon and R. A. Saunders, *Brit. J. Appl. Phys.*, **11**, 128 (1960).

82. 'R. Handley and E. F. G. Herington, *Chem Ind. (London)*, **304** (1956).

83. R. Handley, *Ind. Chem.*, **31**, 535 (1955).

84. E. F. G. Herington, *Research (London)*, **7**, 465 (1954).

85. G. J. Sloan, *J. Am. Chem. Soc.*, **85**, 3899 (1963).

86. M. J. Joncich and D. R. Bailey, *Anal. Chem.*, **32**, 121578 (1960).

87. H. Schildknecht and H. Vetter, *Angew. Chem.*, **73**, 612 (1961).

88. W. D. Betts, J. W. Freeman, and D. McNeil, *J. Appl. Chem.*, **18**, 180 (1968).

89. H. Schildknecht, J. Breiter, and K. Maas, *Separ. Sci.*, **5**, 99 (1970).

90. G. R. Atwood, in *Recent Developments in Separation Science*, N. N. Li, ed., Vol. I, CRC Press, Cleveland, 1972, p. 1.

91. R. S. Tipson, in *Techniques of Organic Chemistry*, A. Weissberger, ed., Vol. 3, Part 1, Interscience, New York, 1956, p. 395.

92. A. Luttringhaus, in *Methoden der Organische Chemie*, J. Houben and T. Weyl, eds., Thieme-Verlag, Stuttgart, 1958, Part 1, p. 341.

93. H. A. Laitinen, *Chemical Analysis*, McGraw-Hill, New York, 1960.

94. G. H. Morrison and H. Freiser, *Solvent Extraction in Analytical Chemistry*, Wiley, New York, 1957.

95. J. Starý, *The Solvent Extraction of Metal Chelates*, Macmillan, New York, 1964.

96. O. G. Koch and G. A. Koch-Dedic, *Handbuch der Spurenanalyse*, Springer-Verlag, Berlin, 1964, pp. 179–270.

97. H. A. Flaschka and A. J. Barnard, Jr., "Complexation in Aqueous Media," in F. D. Snell and C. L. Hilton, eds., *Encyclopedia of Industrial Chemical Analysis*, Interscience, New York, 1966.

98. M. Delepine, *Compt. Rend.*, **146**, 981 (1908).

99. A. E. Martin, *Anal. Chem.*, **25**, 1260 (1953).

100. H. Bode and F. Neumann, *Z. Anal. Chem.*, **172**, 1 (1960).

101. S. Nomoto and F. W. Sunderman, Jr., *Clin. Chem.*, **16** (6), 477 (1970).

102. C. L. Luke, *Anal. Chim. Acta*, **37**, 267 (1967).

103. C. L. Luke, *Anal. Chim. Acta*, **41**, 237 (1968).

104. H. Malissa and E. Schöffmann, *Mikrochim. Acta*, **1**, 187 (1955).

105. W. Slavin, *Atomic Absorption Spectroscopy*, Wiley-Interscience, New York, 1968.

106. J. W. Robinson, P. F. Lott, and A. J. Barnard, Jr., "Flame Analytical Techniques," in H. A. Flaschka and A. J. Barnard, Jr., eds., *Chelates in Analytical Chemistry*, Vol. 4, Dekker, New York, 1972, p. 233.

107. E. F. Joy, N. A. Kershner, and A. J. Barnard, Jr., *Spex Speaker*, **16** (3), 1 (1971).

108. K. Eckschlager, P. Stopka, and J. Veprek-Siska, *Chem. Prum*, **17**, (12), 667 (1967).

109. H. Fischer, *Angew. Chem.*, **50**, 919 (1937).

110. E. B. Sandell, *Colorimetric Determination of Traces of Metals*, 2nd ed., Interscience, New York, 1950.

111. G. Iwantscheff, *Das Dithizon und Seine Anwendung in der Mikro and Spurenanalyse*, Weinheim/Bergstr., Verlag Chemie, 1958.

112. C. C. Patterson and D. Settle, Seventh Materials Research Symposium, National Bureau of Standards, Gaithersburg, Md., October 7–11, 1974.

113. K. L. Cheng and R. H. Bray, *Anal. Chem.*, **27**, 782 (1955).

114. S. Shibata, *2-Pyridylazo Compounds in Analytical Chemistry*, in *Chelates in Analytical Chemistry*, H. A. Flaschka and A. J. Barnard, Jr., eds., Vol. 4, Dekker, New York, 1972, p. 1.

115. A. J. Barnard, E. F. Joy, K. Little, and J. D. Brooks, *Talanta*, **17**, 785 (1970).

116. L. Meites, *Anal. Chem.*, **27** (3), 416 (1955).

117. M. Zief and J. Horvath, *Lab. Prac.*, **23** (4), 175 (1974).

118. W. Matson, Environmental Science Associates, personal communication.

119. G. D. Christian, *Anal. Chem.*, **41** (1), 24-A (1969).

120. M. Zief and A. G. Nesher, *Environ. Sci. Technol.*, **8** (7), 677 (1974).

121. G. A. Carlson, U.S. Patent 3,650,925, March 21, 1972.

122. H. S. Washington, *Chemical Analysis of Rocks*, 4th ed., Wiley, New York, 1930, pp. 220–222.

123. *Ind. Eng. Chem.*, January 28, 1974, p. 24.

124. J. Minczewski, in *Trace Characterization: Chemical and Physical*, W. W. Meinke and B. F. Scribner, eds., National Bureau of Standards Monograph 100, 1967.

125. Z. Marczenko. *Zesz. Nauk. Politech. Warsaw*. **115**; *Chemia* (3) 1965.

126. Z. Marczenko and K. Kasiura, *Chem. Anal.* (*Warsaw*), **10**, 449 (1965).

127. Z. Marczenko and M. Krasiejko, *Chem. Anal. (Warsaw)*, **9**, 291 (1964).

128. E. C. Kuehner and R. T. Leslie, in *Encyclopedia of Industrial Chemical Analysis*, F. D. Snell and C. L. Hilton, eds., Vol. 3, Interscience, New York, 1966, p. 572.

129. C. A. Holden and H. S. Bryant, *Separ. Sci.*, **4** (1), 1 (1969).

130. R. D. Mounts in Ref. 12, p. 101.

131. M. Tatsumoto and J. N. Rosholt, *Science*, **167**, 461 (1970).

132. M. Zief, Symposium Sponsored by Committee on Specifications and Criteria for Biochemical Compounds, National Academy of Sciences, National Research Council, San Francisco, June 17, 1971.

133. *Matheson Gas Data Book*, 4th ed., Matheson Co., East Rutherford, N.J., 1966.

134. *Handbook of Compressed Gases*, Compressed Gas Assoc., Reinhold, New York, 1966.

135. C. A. McMenamy, personal communication, 1975.

136. L. Meites, *Polarographic Techniques*, 2nd ed., Wiley-Interscience, New York, 1965, p. 341.

137. W. Daniel and H. Schwedler, *Z. Anal. Chem.*, **99**, 385 (1934).

138. J. Falbe and H.-D. Haln, M. M. Wright, *Anal. Chem.*, **26**, 1001 (1954).

139. M. P. Jacobs, M. M. Braverman, and S. Hockheiser, *Anal. Chem.*, **29**, 1349 (1957).

140. R. S. Fielder, P. J. Jackson, and E. Raask, *J. Inst. Fuel*, **33**, 84 (1960).

141. H. G. Eaton, M. E. Umstead, and W. D. Smith, *J. Chromatogr. Sci.*, **11**, 275 (1973).

142. F. W. Williams, F. J. Woods, and M. E. Umstead, *J. Chromatogr. Sci.*, **10**, 570 (1972).

143. C. W. Skarstrom, U.S. Patent 2,944,627.

144. R. H. Johns, *Chem. Eng. Progr. Symp. Ser.*, **62**, 81 (1966).

145. J. K. Musick, F. S. Thomas, and J. E. Johnson, *Ind. Eng. Chem., Process Design Develop*, **11**, 350 (1972).

146. F. W. Williams and H. G. Eaton, *Anal. Chem.*, **46**, 179 (1974).

147. C. K. Hersh, *Molecular Sieves*, Reinhold, New York, 1961.

148. J. Falbe and H. Hahn, in *Methodicum Chimicum*, F. Korte, ed., Academic Press, New York, Vol. 1 (B), 1974 p. 990.

CONTAMINATION CONTROL DURING ROUTINE ANALYTICAL OPERATIONS

The recent analytical literature reveals that analysts increasingly prefer to adopt sensitive, instrumental methods for determining trace elements. Such methods are advantageous when they can be performed nondestructively, since contamination problems are diminished. However this approach is not universal inasmuch as the ideal sample directly amenable to nondestructive techniques without interferences occurs infrequently. In the real world, a majority of samples must be treated chemically or physically to produce a species suitable for determination. This may include chemical treatment to enhance the selectivity and sensitivity of the method either by separating matrix elements or by concentrating trace impurities. Mechanical treatment to avoid problems of segregation may also be required. It is therefore necessary to frequently use routine procedures such as sampling, weighing, measuring, drying, heating, fusion, ashing, dissolution, stirring, concentration, evaporation, and filtration. In ultratrace analysis these procedures must be rigidly controlled, since high susceptibility to contamination exists.

Contamination of samples during routine operations does not significantly influence analytical results at concentrations above 100 μg/g. For valid data below this level, routine operations must be performed with meticulous care. This chapter treats the common sources of contamination and describes techniques for performing standard operations to minimize the introduction of systematic errors due to the blank.

I. SAMPLING

A. SOLUTIONS AND LIQUIDS

Solutions or liquid reagents contained in small quartz or Teflon bottles are conveniently sampled and measured with quartz utensils or with disposable plastic pipets. The exterior of the container is cleaned with nitric acid, then rinsed with ultrapure water and air dried in a laminar-flow hood. The chemist, wearing a good grade of polyvinylchloride glove, opens the bottle in the clean hood. The liner in the bottle cap is inspected to detect signs of chemical reaction with the contained liquid. Containers for high quality chemicals should have virgin (nonpigmented) polyfluorocarbon or polyethylene caps fitted with Teflon liners. Highly colored caps or those having paper liners usually contain a variety of traces at concentrations above 0.5 $\mu g/g$ and will seriously contaminate any contained reagent.

As the opened container is slowly rotated, a portion of liquid is poured directly into a disposable polyethylene beaker until the entire lip of the bottle has been rinsed thoroughly with solution. This portion is discarded, and a second small volume is poured directly from the bottle into another clean polyethylene beaker. After rinsing the beaker with this sample, a third portion of the reagent is poured into the beaker and is then sampled with a quartz or disposable plastic syringe or pipet.*

Liquids can be contaminated in several ways during the sampling process. Pigments used to indicate volume increments on plastic or glass measuring devices can impart impurities. Suction bulbs frequently release particles of rubber, which may contain the following elements added as accelerators, stabilizers, and catalysts: Al, Ba, Cd, Co, Cu, Fe, Hg, Mo, Pb, Sb, Sn, Ti, and Zn. These particles also readily contaminate organic solutions with leachable components of the rubber.[1] Disposable plastic pipets with cotton plugs prevent particle fallout of rubber from suction bulbs. To avoid the introduction of organic impurities, solutions must not be allowed to contact the cotton.

Sampling of natural waters requires extreme care to minimize the introduction of impurities from collection devices. In addition, pH adjustment of the sample is necessary to prevent losses of trace elements by adsorption on container walls. Robertson suggests that the trace elemental abundances found in seawater have been lowered as sampling and analytical methods have become more refined.[1] Present-day methods for sampling the contents of large bodies of liquids—for example, steam boilers, storage tanks, sewage waste and process streams—could benefit from the wealth of information available on obtaining and storing representative samples of natural waters.[2-4]

B.   POWDERS AND SOLIDS

Granular or powdered substances can be sampled with polyethylene tubing that is cut at an angle on one end. Teflon-coated or quartz spatulas are also convenient noncontaminating tools for transferring small samples that have been poured directly from reagent bottles into plastic beakers. Spatulas of easily abraded materials must not be used for hard granular samples, since particulate contamination can result.

Bulk solids present the most difficult sampling problem because crushing

---

* See section on measuring solutions in chapter 2 for details.

and grinding entail several difficulties. Contamination introduced by grinding is difficult to assess by a blank determination, since constant blanks during repetitive grinding operations are highly unlikely. Impurities from the grinding vessel may also interfere with the detection of traces in the sample. If grinding is to be performed, the first step is to select a mortar constructed from a material much harder than the sample. This material should also be free from impurities undergoing measurement. Common materials for fabricating mortars and pestles are listed in Table 1.[5, 6] Except for extremely abrasion-resistant materials such as tungsten carbide and boron carbide, contamination during grinding can be expected.

Analysis of the bulk solid or a solution of the solid is preferable to pulverization, but unfortunately this is not always possible. For example, the preparation of certain standards and samples for emission spectrographic work or mass spectrometric analyses of solids, requires grinding and mixing of the sample with graphite. On the other hand, grinding should not be employed to enhance the dissolution of samples or to obtain small samples for weighing. Alternative ways of increasing the dissolution of samples without pulverization are discussed extensively later in this chapter. Brittle solids can be broken into several coarse pieces by wrapping in polyethylene film, placing between thin sheets of Teflon (TFE) and crushing with blows from a hammer or by means of a hydraulic press. A sample of appropriate weight may then be selected.

When grinding is unavoidable, the following precautions will minimize

TABLE 1. Materials for Mortars, Pestles, and Grinding Vials

Material[a]	Contaminants Introduced	Samples Tested	Reference
Tungsten carbide	Co, Ti	SiO_2, $CaCO_3$	6
Alumina	Al, Cr, Fe, Ga, Zr	SiO_2, $CaCO_3$	5
Alumina-ceramic	Al, Cu, Fe, Ga, Li, Ti, B, Ba, Co, Mn, Zn, Zr	SiO_2, $CaCO_3$	5
Boron carbide	B, other traces not detected	SiO_2, $CaCO_3$	5
Lucite (polymethyl methacrylate)	No traces detected	SiO_2, $CaCO_3$	5
Silicon	Cu, Ca, Al, Mg, Fe, Ni	Silicon	7
Agate	Cu, Ca, Al, Mg, Fe, Ni	Silicon	7
Quartz	Cu, Ca, Al, Mg, Fe, Ni	Silicon	7
Plattner mortar	Fe, Mn, Cr, Ni, Cu	Quartz	8
Molybdenum	1% Mo	Silicon	9

[a] Nanograms per gram of impurities introduced by grinding four materials. *Silicon*: Cu, 63; Ca, 710; Al, 1100; Mg, 950; Fe, 630; Ni, 110. *Agate*: Cu, 131; Ca, 1600; Al, 2200; Mg, 1600; Fe, 710; Ni, 140. *Quartz*: Cu, 74; Ca, 3100; Al, 4400; Mg, 10; Fe, —; Ni, 830. *Plattner mortar*: Fe, 2.8×10^4; Mn, 1.8×10^3; Cr, 4.0×10^2; Ni, 2.5×10^2; Cu, 3.5×10^2.

the introduction of contaminants: (1) one mortar and pestle should be reserved for a particular material to prevent cross-contamination, (2) pre-grinding an aliquot of the sample and discarding it before the main sample is treated diminishes contamination from the mortar, and (3) if survey analyses are to be performed, several mortars characterized for trace element content should be used. Grinding by hand with a mortar and pestle has been reported to be the safest method.[7]

Hard materials or large metal rods can be cut with a diamond-tipped saw or dissected by exposure to a laser beam. Although the latter method has not yet been developed into a convenient laboratory technique, it promises to be a most helpful tool in preventing contamination during the preliminary sampling of hard solids. When trace elements are to be measured in samples sectioned with laser beams, it will be necessary to investigate the effects of selective vaporization or migration of traces in the region of the sample heated by exposure to the beam. Vaporization of material and its redeposition upon cooling on the surface of the sample, and spontaneous fracture of the sample due to thermal stress, are several envisioned problems that will need to be resolved for certain applications. The utilization of this tool in sampling should be explored fully by analytical chemists.

After mechanical treatment of bulk solids by pulverizing, fracture, or cutting, appropriate cleaning procedures are necessary to remove surface contaminants. Preliminary washing of semiconductor grade bulk silicon with aqua regia and water decreases the impurity content of the sample.[8] Other comparisons between different methods of grinding followed by washing with aqua regia and high quality water show the importance of cleaning the surface prior to dissolving the sample.[8] In this laboratory γ-ray spectrometry has been employed to verify the need for cleaning surfaces of irradiated samples before dissolving the sample for counting.[9] Etching samples with ultrapure acids, rinsing with high purity water, and drying under heat lamps in a laminar-flow hood is recommended.

Surface cleaning techniques based on sputtering or bombarding with noble gas ions may also be employed, but these specialized techniques are seldom applicable for cleaning samples to be analyzed by conventional methods; when these methods are applicable, however, they are highly effective. For example, high purity tantalum plates were presparked in a mass spectrometer to remove any residual surface impurities that remained after cleaning the plates in HNO_3 and distilled water. Samples of the solution to be analyzed were then evaporated onto the plate, sparked, and compared with a doped sample and with a presparked blank plate to obtain semiquantitative determinations of many trace elements.[10]

II. HEATING AND DRYING

A. GAS BURNERS AND HOT PLATES

Conventional methods of heating are not suitable for ultratrace analysis. Metallic gas burners and unprotected metallic heating coils must be avoided. Since metallic parts of hot plates are attacked readily by acids and emit contaminating particles, ceramic-top hot plates or heating coils encased in quartz are preferable.

B. FURNACES

Conventional muffle furnaces that have aged are often sources of serious contamination. High temperature ceramic liners are usually responsible for trace aluminum, zirconium, or chromium contamination.[11] Chromium in processed silicon wafers, for example, was traced to the heating elements and liner of the diffusion oven.[12]

Furnaces lined with quartz sleeves have been used for low temperature (< 1200°C) heating, fusion, or drying. The apparatus in Figure 1 was spe-

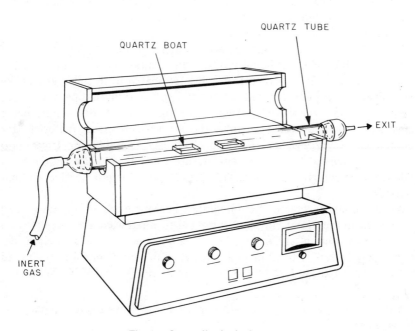

Fig. 1. Quartz-lined tube furnace.

cially fabricated for drying ultrapure compounds for subsequent analysis. Here water vapor is swept from the chamber by the application of a continuous positive pressure of nitrogen or argon. A 47 mm in-line Swinnex* holder containing a 0.2 μm cellulose acetate membrane filter eliminates particulate matter. Substitution of quartz or Teflon boats (at lower temperature) for platinum vessels is usually necessary to eliminate trace contamination from the container during the drying process. The exit tube connected by a standard taper joint is removed for loading and unloading the boats. On these occasions the exit tube must always be positioned in a laminar-flow hood, to expose the open quartz tube to inert gas from one direction and to class 100 air from the clean hood. The boats are removed from the tube via a long 6.4 mm quartz rod bent at one end to slip over a handle on the boats. The entire furnace is mounted on a table provided with polyvinyl chloride casters, allowing the analyst to move it into position during loading and unloading. Quartz supports for the boats during unloading at elevated temperatures are custom built to the correct height.

C.   MICROWAVE HEATING

Significant advantages are offered by microwave heating devices for removing water from solids,[13] for drying plastic containers,[14] and for flash evaporating liquids.[15] A household commercial oven† has been used by one of the authors for water removal in trace analysis and ultrapurification of inorganic solids. Solids, liquids, or plastic containers were placed in a Halar‡ tray (28 × 28 × 10 cm) specially fabricated to fit the cavity of the oven. The tray was covered with a loose-fitting Halar lid that permitted free passage of water vapor and prevented the spattering of dry solids. Being transparent to microwave energy, plastics are preferred materials for containers. Polypropylene, Halar, or Teflon is suitable for trays; polyethylene softens at too low a temperature for extended use. For neutral and acidic chemicals, quartz dishes provide the longest service without breakage.

When sodium carbonate monohydrate (theoretical water content 14.5%) was placed in the Halar tray and subjected to microwave energy, the solid became anhydrous after 20 min. Moisture was determined by weight loss after drying at 130°C for 2 hr in a conventional oven. Control tests showed that the monohydrate lost 14.45% moisture after 1 hr at 130°C. When 100

---

* Millipore Corp., Bedford, Mass.

† Litton household microwave oven, Model 402,001, output 350 W, Litton Industries, Minneapolis, Minn.

‡ Halar is a 1:1 alternating copolymer of ethylene and chlorotrifluoroethylene, Allied Chemical Co., Morristown, N.J.

ml of a 25% aqueous solution of sodium carbonate was placed in the tray, the water was substantially removed in 13 min. The residual solid was ground to a fine powder with a quartz mortar and pestle, then returned to the microwave oven for another 10 min to prepare an anhydrous product. Polyethylene and polypropylene bottles cleaned by leaching with concentrated nitric acid and rinsing with deionized water were dried within 15 min in a microwave oven.

A microwave oven operating with a frequency of 2450 MHz and an output of 350 W on 110 V will find many applications in trace analysis. (See Standards Section IV, Chapter 2). When a more powerful oven with the same microwave frequency operating on 220 V with 1300 W output was used to concentrate a sodium carbonate solution, intense corona discharges throughout the oven cavity were observed. Recently a commercial microwave oven for general laboratory use was announced.* The unit operates on 110 V at 60 Hz and features high and low power levels, a digital readout timer, and adapters permitting purging with inert gas.

D. AIR DRYING AND INFRARED HEATING

Freedom from airborne particulates during drying is eliminated by laminar-flow hoods or closed plastic chambers that are purged continuously with filtered air. Plastic or glass items are conveniently dried on polypropylene racks in clean hoods under infrared lamps as in Figure 2.

The assembly outlined in Figure 3 was suggested by Thiers in the late 1950s.[8] This approach is still appropriate today for eliminating contamination while drying solids or concentrating liquids. A glass as well as a Teflon assembly with a polyethylene cap (Figure 4) has been used successfully at the Bell Laboratories. Hydrolyzed $SiCl_4$ samples have been dried in these assemblies, and trace elements were not detected in blanks obtained by exposing empty containers under identical conditions.

III. FUSION

A. CONVENTIONAL METHODS

Fusions with acidic or basic fluxes must be carefully evaluated in ultratrace analysis. Because of impurities in the flux and contamination from the crucible, the blank value can seldom be reduced to a level that permits reliable determination of submicrogram traces. Unless new developments in the ultrapurification of fluxes and novel noncontaminating

* Sage Laboratories, Inc., Natick, Mass.

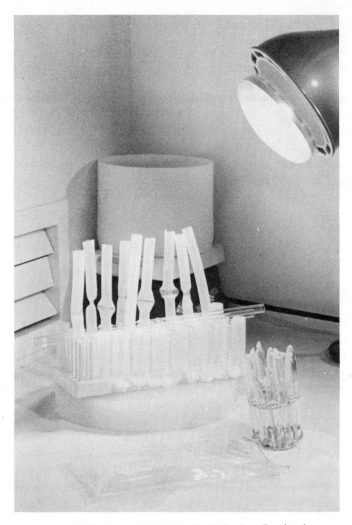

Fig. 2.   Drying by infrared heating in laminar-flow hood.

materials for crucibles become available, fusion will not be applicable for treating samples in which ultratrace elements are to be determined.

Several precautions are recommended in the high temperature melting of fluxes. Radiofrequency induction heating in crucibles constructed from materials that introduce only tolerated contaminants has been used advantageously. The system in Figure 5 has been used for melting high

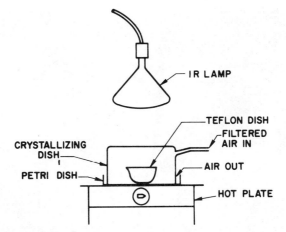

Fig. 3. Chamber for evaporation (after Thiers). Reprinted by permission from *Methods of Biochemical Analysis,* Vol. 5, D. Glick, ed., Wiley, New York, 1957, p. 279.

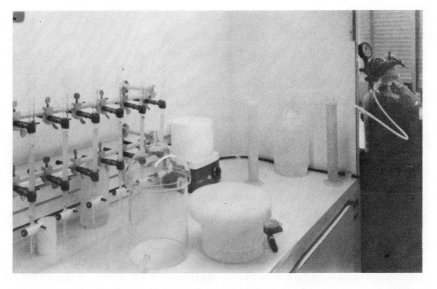

Fig. 4. Teflon and borosilicate glass chambers.

Fig. 5. Radiofrequency inducion melting in laminar-flow hood. *Source*: Ref. 16. Copyright 1972, Bell Telephone Laboratories, Incorporated. Reprinted by permission, Editor, Bell Laboratories RECORD.

purity glasses.[16] Premelting of the raw materials is confined to a laminar-flow hood to prevent airborne particulate contamination. Alternatively, protection from dust during melting can be provided via the system in Figure 6, which affords protection for the contents of the container without enclosing the entire furnace in a controlled atmosphere environment.[16]

Platinum crucibles have been widely used for high temperature fusions, but contamination of the melt by platinum at the microgram per gram level usually occurs. Platinum, iridium, gold and iron have been detected easily in melts and in glasses fused in platinum vessels.[17] Emission spectrographic survey analyses have demonstrated that platinum contains a variety of trace impurities as given in Table 2. These data show that the level of contamination by some traces in platinum decreases as the crucible is reused.

The amount of platinum that contaminates melts has been found to be less when fusions are performed under nitrogen atmospheres than when melting is done in oxygen.[17] Although prefusion of high temperature fluxes and controlled atmospheres can be used to reduce contamination by platinum vessels, replacement of platinum with other more suitable materials is recommended. Since iridium has a much higher usable temperature range, up to 2300°C compared to less than 1600°C for

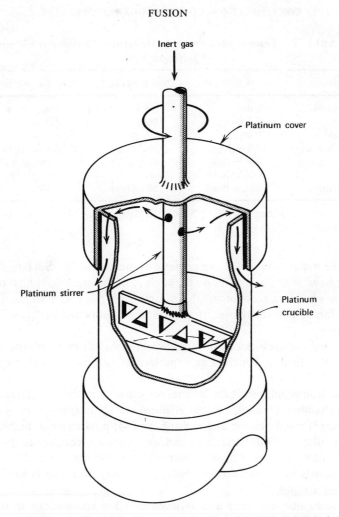

Fig. 6. Gas-purged platinum crucible. *Source*: Ref. 16. Copyright 1972, Bell Telephone Laboratories, Incorporated. Reprinted by permission, Editor, Bell Laboratories RECORD.

platinum, its substitution could result in less container contamination. When the usable temperature ranges of materials are comparable, resistance to attack by the melt is the most important parameter to be considered.

B. LEVITATION MELTING

In processing extremely pure, high melting point materials, for example, silicon, germanium, titanium, or zirconium, contamination is unavoid-

TABLE 2.   Trace Impurities (wt %) Detected in Platinum by Emission Spectrography[a]

Element	New Crucible 1	New Crucible 2	Crucible (control)
Copper	0.000x	0.000x	0.00x
Iron	0.000x	0.000x	0.000x
Gold	0.00x	0.00x (low)	0.000x
Palladium	0.00x (high)	0.00x	0.000x (high)
Rhodium	0.000x (high)	0.000x (high)	0.000x (low)
Silver	0.000x (low)	0.000x	0.000x (low)
Titanium	0.000x (low)	0.000x (low)	0.000x

[a] Qualitative analysis by D. L. Nash, Bell Laboratories, Murray Hill, N.J.: x (high) = 5–9, x (low) = 1–4.

able if the material is heated in a ceramic or metal crucible. Such problems as the release of water and gases on the surface of the crucible at high temperature, the chemical reaction of the sample with the container, and mutual dissolution of both at the interface, are only eliminated by "crucible-less" techniques.

Electrical conductors can be suspended (levitated) in an alternating electromagnetic field and melted or vaporized at sufficiently high frequencies (10–500 kHz). The main advantages of levitation melting include (1) freedom from contamination by sources other than the atmosphere in the sample chamber, (2) simultaneous efficient electromagnetic stirring by eddy currents, (3) rapid heating and melting, and (4) production of highly homogeneous alloys.[18] Difficulties can include expensive equipment, processing only of small amounts of highly electrically conducting materials, instability of and difficulty in measuring temperature, and excessive evaporation or fuming of samples.[19]

The technique has been mainly developed for applications in materials production rather than as an analytical tool. Extremely pure silicon crystals were prepared at Standard Telecommunication Laboratories by fabricating cold crucibles in which the silicon charge was levitated a small distance above the surface of the water-cooled crucible. Niobium, molybdenum, tantalum, and rare earths have been purified in this way.[18] At Bell Laboratories E. Greiner and E. Buehler used the cold crucible method to prepare rods of intermetallic compounds such as V_3Si, V_3Ge, and Nb_3Pt.[18]

In analytical chemistry the technique can best be applied to the production of small batches of homogeneously doped metals for use as standards. Oxygen and nitrogen in ferrochromium have been homogeneously dis-

TABLE 3. Multiple Analyses of a Sample of Ferrochromium[18]

Amounts (ppm) Before Levitation Melting		Amounts (ppm) After Levitation Melting	
O_2	N_2	O_2	N_2
476	389	751	228
450	368	761	226
408	360	764	230
406	339	760	220

tributed in the alloy by levitation melting in the respective atmospheres (Table 3). The list of metals successfully levitation melted (Table 4) indicates that there can exist a variety of high purity metals in which gases are homogeneously distributed; alloys for neutron flux monitors can also be prepared.

Harris and co-workers[20, 21] reported a vacuum fusion gas analysis method based on levitation melting. The metal sample to be analyzed for oxygen was introduced into a levitated liquid sphere of iron containing carbon, and the evolved carbon monoxide was determined within 10 min.

To increase the sensitivity of methods for determining nitrogen in refractory metals such as niobium, a new decomposition procedure was developed in which the sample was levitation melted.[22] In this procedure the blank was reduced because no sample holder was used, and it was possible to differentiate between absorbed surface nitrogen and nitrogen within the sample. Surface nitrogen was baked out by stepwise heating, and then dissolved nitrogen was released upon fusion of the sample at higher temperatures. A mass spectrometer was used to measure the released nitrogen. Other innovative uses of levitation melting remain to be exploited by analytical chemists.

TABLE 4. Elements Successfully Levitation Melted

Ag	Cr	Ga	Li	Pb	Si	Y
Al	Cu	Gd	Mg	Pd	Sm	Yb
Au	Dy	Ge	Mo	Pr	Sn	
Be	Er	Ho	Nd	Ru	Ti	
Co	Fe	In	Ni	Sb	V	

Source: Ref. 18.

IV. ASHING

A. WET AND DRY ASHING

To destroy organic matter while dissolving stubborn matrices such as bone, teeth, and synthetic polymers, acid digestion at elevated temperatures is required. This procedure is plagued by contamination resulting from impure acids, the containers, and the furnace or hot plates used for heating. Another problem, losses of trace elements by volatilization during wet ashing, is less serious than it is during dry ashing.[23, 24] Except for impurities introduced with the acids, dry ashing procedures are subject to the same hazards as wet ashing. Additional problems encountered in simultaneously ashing filter or weighing paper with the sample must be considered. During dry ashing at temperatures in excess of 400°C, serious losses of trace elements are caused by diffusion, volatilization, and adsorption. Contamination by the container can also be significant.

During the oxidation of organics in dry ashing, the temperature in portions of the sample can be elevated well above the furnace temperature. This local heating, which increases losses of traces by volatilization, can be prevented by preashing. In addition, the analyst must compensate for the formation of insoluble compounds, the strong adsorption of cations onto silica or porcelain dishes, and the presence of adsorbed metals from previous ashings in a crucible.

The primary problems inherent in dry ashing procedures, listed below, were previously reviewed by Thiers.[25]

1. Unduly high temperature of the sample.
2. Incomplete ashing resulting in small amounts of carbon residues.
3. Use of etched glass, ceramic, or silica dishes.
4. Large concentrations of silica in the sample.
5. Prevailing acid conditions during ashing.

B. LOW TEMPERATURE ASHING

Decomposition of organic substances in a stream of highly reactive oxygen, produced by a high frequency electromagnetic field, is a recommended alternative to conventional high temperature dry or wet ashing procedures. In this method electrical energy is transferred directly to a stream of low pressure gas, producing highly excited states of oxygen that react selectively with elements in the organic sample. With this method, complete destruction of biological samples occurs at 67°C.[26] Because general heating is minimized, the sample temperature is low and loss of trace elements by volatilization is substantially decreased. Investigations of volatility losses of radioisotopes in blood samples showed 99 to 102% recovery of Sb, As, Cs, Co, Cr, Fe, Pb, Mn, Mo, Se, Na, and Zn.[27] However appreciable losses of Ag, Hg, I, and Au were observed. Quantitative recoveries without container contamination of submicrogram amounts of boron from internal organs of rats have also been accomplished by this technique.[28] A commercial unit was recently made available.*

V. DECOMPOSITION AND DISSOLUTION

A. ACID DIGESTION

Although ordinary methods for dissolving samples by acid digestion or by fusion are seldom acceptable for ultratrace analysis, several simple precautions can be taken to minimize sample contamination. Samples should be treated with the minimum quantity of high purity solvent or acid and heated in one of the appropriate gas-purged chambers described previously.

Perhaps the single most damaging factor in procedures for dissolution is the purity of the reagents used to digest the sample. An acid with 1.0 ng/ml of a cation could contribute 10–50 ng of this trace during repetitive additions of acid to ensure complete dissolution of the sample. Since the blank value should ideally be less than one-tenth the amount of the element in the sample, the lower limit for the determination may be on the order of 100–500 ng. Therefore one must first obtain the purest acid possible, and use the smallest volume necessary for dissolution.

B. ACID PRESSURE BOMBS

The merits of acid pressure decomposition techniques for dissolving samples are indeed significant in quantitative trace analysis.[29] Rapid dissolution of organics and refractory inorganics can be achieved. For example, the determination of mercury in fish requires digestion for 3 to 4 hr in a mixture of sulfuric, nitric, and perchloric acids; but dissolution by pressure decomposition requires only 30 min.[30] Equally impressive reductions of time have been found in our laboratory for the solubilization of inorganics. A list of the shortcomings of wet ashing has been compiled by Bernas.[30] Pressure decomposition overcomes these drawbacks, which are given below.

1. Incomplete mineralization of substances that are especially difficult to decompose in open vessels at relatively low temperatures.

2. Difficulty in obtaining complete extraction of the metal being determined from certain ignited residues.

* International Plasma Corp., Hayward, Calif.

3.   Prolonged boiling leading to charring, which could hold back some trace metal by absorption or occlusion.

4.   Excessive charring that leads to losses, particularly in the presence of chlorides.

5.   The ever-present risk of volatilization and retention losses—the latter due to the combination of the element to be determined with the material of the container, be it metal or glass.

6.   Contamination by container and environment.

7.   Wet methods failing to give good recoveries at low concentration levels.

8.   Sources of incomplete recoveries: present-day procedures for destruction of organic matter.

9.   Lack of precision due to difficulties in the destruction of organic matter and isolation of the metal, rather than difficulties in the determination.

10.   The methods require considerable supervision.

11.   The methods are time-consuming and involve hazards in handling hot concentrated acids.

Losses of traces by volatilization are virtually eliminated during pressure decomposition. After reaction of the sample with high purity mineral acids at 110–170°C, the closed vessel is allowed to cool to room temperature before opening. Most cations are quantitatively retained.[29]

A variety of Teflon-lined bombs have been constructed.[31-33] Most are similar to the schematic diagram in Figure 7. Variations involve the size of the Teflon liner,[29] inclusion of a Teflon spout to aid sample transfer,* and different supporting metal jackets of aluminum† or steel. Most commercial devices are designed for a single sample. One of the authors designed and constructed a unit capable of simultaneously decomposing six samples (Figure 8). The diagram for the single device indicates that the Teflon chamber was not completely enclosed in a metal holder (Figure 9). This design prevented acid vapors from contacting metal when the bomb was opened before being cooled completely to room temperature.

Numerous applications of pressure dissolution have been reported.[24, 34, 35] When ultrapure acids are available, pressure dissolution in Teflon devices is highly recommended.

C. GAS PHASE REACTIONS

The ultimate approach to preserve sample integrity is the exploitation of gas phase reactions. When the matrix can be separated efficiently by

* Uni-Seal Decomposition Vessels Ltd., P.O. Box 9463, Haifa, Israel.
† Parr Instrument Co., 211 53rd St., Moline, Ill.

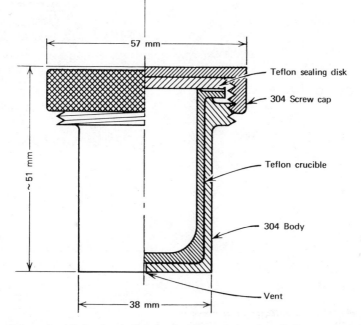

Fig. 7. Schematic of Teflon bomb. Reprinted with permissions from B. Bernas, *Anal. Chem.,* **40,** 1683 (1968). Copyright by the American Chemical Society.

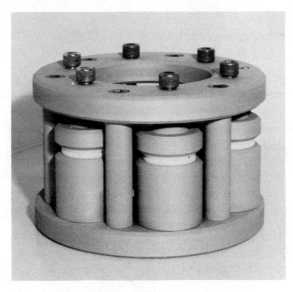

Fig. 8. Teflon bomb cluster apparatus.

167

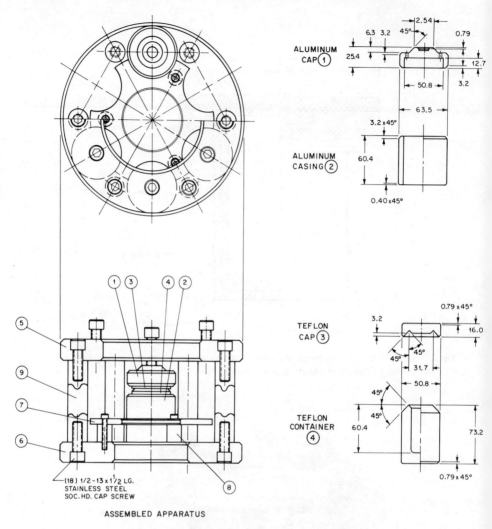

Fig. 9. Schematic of Teflon bomb. Reprinted with permission from J. W. Mitchell, *Anal. Chem.*, **45**, 497A (1973). Copyright by the American Chemical Society.

volatilization or when reactions selectively and quantitatively evolve gaseous products of trace elements, such reactions can be used most effectively to eliminate contamination.

A Teflon apparatus, in which vapors of HF could be generated, was designed specifically to eliminate contamination during the destruction of samples of ultrapure silicon, fused silica, quartz and glass.[36] Figures 10 and 11 illustrate all the parts of the apparatus that were machined at Bell Laboratories from virgin Teflon (TFE) stock. The reagent reservoir D makes a

Fig. 10. Teflon apparatus for vapor phase destruction of silicates.

pressure seal with a groove, machined on the bottom of vapor chamber C. The doughnut-shaped sample rack B then fits snugly onto the chimney of the vapor chamber and is positioned by depressing the plate until it rests on the mantle at the bottom of the chamber. Up to eight Teflon beakers (39 mm diameter × 48 mm) can be placed on the rack. After the chamber cap, machined to fit the top lip of the vapor chamber, is positioned, the assembled unit is supported and sealed tightly by enclosing it in an aluminum framework. The assembled apparatus appears in Figure 12.

The vapor chamber and reagent reservoir are jacketed with custom-fitted, silicone rubber coated heating blankets.* The heating units are equipped with a thermostatic regulator for controlling temperature to ±5°C. These

* Watlow Electric Co., St. Louis, Mo.

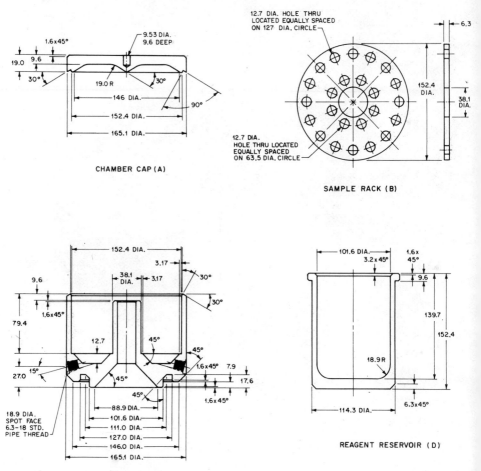

Fig. 11. Schematic diagram of Teflon vapor chamber. Reprinted with permission from J. W. Mitchell, *Anal. Chem.*, **45,** 497A (1973). Copyright by the American Chemical Society.

heating units are supplied with 60 Hz, 120 V current and regulated via an appropriate Variac. The heaters supply a maximum of 0.52 W/cm² of power for heating chamber C and reservoir D.

After the apparatus is assembled in a laminar-flow clean hood and placed in the aluminum housing, 500 ml of HF (ULTREX® grade* or isopiestic distillation of 50% reagent) are carefully poured into the reagent reservoir

* J. T. Baker Chemical Co., Phillipsburg, N.J.

Fig. 12.   Teflon vapor chamber assembly. Reprinted with permission from J. W. Mitchell, *Anal. Chem.*, **45**, 497A (1973). Copyright by the American Chemical Society.

through a polyethylene funnel inserted into the chimney of the vapor chamber. The samples are weighed into 50 ml Teflon beakers and placed on the perforated sample rack.

One to 2 g of bulk, fused silica, silicon, silicon dioxide, and glass, or up to 10 g of these materials in granular or powdered form can be treated. Crushing or grinding samples before treatment should not be attempted. Fracture of bulk samples into smaller pieces increases the efficiency of decomposition and is recommended only when followed by appropriate cleaning to remove surface contaminants.

After the apparatus is capped and sealed, it is transferred to a laminar-flow hood containing an exhaust vent or placed in a conventional exhaust hood. The vapor chamber is then heated to 160°C and allowed to equilibrate for 30 min before the regulator of the heating blanket on the reagent reservoir is turned to a setting that maintains the temperature at approximately 110°C.

By operating in this manner no HF or water condenses in the sample chamber. Excess vapor exits from the chamber via a Teflon (FEP) tube, attached to an adapter and connected to a polyethylene water reservoir.

Samples of granular silicon dioxide weighing 5 g require 10 hr of continuous exposure, whereas 1 g bulk samples require up to 24 hr of treatment.

The operation is terminated by removing the exit tube from the water reservoir and turning off the power to the reagent reservoir. After 30 min the power to the vapor chamber is stopped.

This apparatus offers several advantages for the destruction of silicate matrices. The closed system is continuously purged by a positive pressure of HF vapor during operation. By using vapors from hydrofluoric acid (an azeotropic mixture containing 35.6% HF, 64.4% H_2O at 111.4°C) trace impurities in the liquid reagent are not added to the sample. Thus blank values have been observed to be considerably lower than those occurring during the dissolution of silicates in liquid HF, where trace contaminants in the reagent are concentrated into the sample during repetitive boiling and fuming with fresh portions of HF.

Reagent grade HF can be used, since the acid is purified during vaporization. In this case violent boiling in the reagent reservoir must be eliminated to prevent liquid HF in the form of small droplets or fine spray from being transferred into the vapor chamber.

Recent experiments have shown that silicon dioxide in a Teflon beaker can be decomposed at 150°C by reaction with vapors of HF produced in a conventional Teflon bomb. The sample in this case is placed on a perforated plate and suspended over HF in the bomb. This simple inexpensive apparatus is quite convenient for treating a single sample and can be easily made by modifying an available Teflon bomb.

Matrices other than silicon can also be effectively treated by gas phase reactions. Vaporized bromine trifluoride has been used to decompose tungsten.[37] Alkali metals in aluminum were determined after volatilization of aluminum as an organometallic compound.[38] Gallium has been decomposed by reaction at 600°C with chlorine.[39] Tungsten, molybdenum, and germanium have been allowed to react with CCl_4 at 900°C to determine nonvolatile chlorides.[40-42] Matrix elements may also be removed as volatile oxides of boron, carbon, selenium, rhenium, osmium, and ruthenium, and as chlorides of silicon, germanium, selenium, tellurium, phosphorus, arsenic, antimony, and tin. Trace elements, especially mercury, have been effectively separated in the elemental state. Successful separations have also been based on the use of volatile compounds ($GeCl_4$, $AsCl_3$, SeO_2, CrO_2Cl_2,

As_2O_3), bromides (Sn, As, Sb, Se, and Re), oxides (OsO_4, Mn_2O_7), and also gaseous compounds (BF_3, SiF_4, H_2S, H_2Se, CO_2, AsH_3, PH_3, and NH_3).[43]

Vapor phase dissolution of other ultrapure materials or samples containing traces should afford procedures far superior to careful dissolution in liquids. For example, organic materials generally should be completely broken down by mixed vapors of HNO_3-$HClO_4$ or HNO_3-H_2SO_4. The potential of vapor phase reactions for analytical purposes has not yet been fully exploited by analytical chemists.

VI. CONCENTRATION OF TRACE ELEMENTS

Traces of inorganic elements in large volumes of solutions must often be preconcentrated before determination by many analytical techniques. Concentration, of course, is a risky operation that should be avoided whenever possible by using methods sufficiently sensitive for direct analysis. This does not imply that careful techniques cannot be executed without reducing the susceptibility to contamination problems.

A. ELECTROLYSIS

Fundamental aspects of the electrochemistry of dilute solutions and practical considerations of the quantitative electroseparation of submicrogram amounts of elements were discussed as early as 1946 and 1950, respectively.[44, 45]

The importance of the supporting electrolyte in speeding deposition has been well known, and the electrolyte recognized as a source of contamination. Equipment for purifying such chemicals via mercury cathode electrolysis is now available (see Electrolysis, Section I.F, Chapter 5). Serious problems are presented by negative losses by adsorption of traces on the walls of the electrolytic cell, at the ends of salt bridges, and on electrodes. For cations at submicrogram concentrations these processes may be sufficiently strong to compete successfully with electrodeposition.[46]

Losses by adsorption can be decreased considerably if the element is first complexed in solution, then electrolyzed. For example, the appreciable adsorption of Ag^+ from nitrate and perchlorate solution was considerably reduced in ammoniacal solution and completely eliminated in cyanide media.[47] Quantitative deposition of some cations are impossible because of colloidal formation and the solubilization of certain compounds (oxides, etc.) at extremely low concentrations.[48] Rogers observed that up to 20% of submicrogram quantities of trace elements deposited on inert electrodes can be lost by dissolution when the electrodes are rinsed.[46]

Small amounts of reducible metal ions can be concentrated from large volumes of dilute solutions by electrodeposition onto pyrolytic graphite electrodes.[49] For subsequent X-ray fluorescence analysis, a thin disk is cleaved from the electrode. Advantages of this system are reported to include graphite disks that are durable, easily stored, and ultrapure with a minimum of background interference.[40, 47] As Table 5 indicates, 6–40 μg of copper, mercury, zinc, nickel, cobalt, and chromium ions have been deposited onto these electrodes, with the recoveries given.

Electrode assemblies have been devised for electrically depositing copper

TABLE 5. Electrodeposition[a] of Six Elements in 0.033 M K₂SO₄

Metal	Known Concentration (μg)	Number of Runs	Amount Found (μg)	S.D. (μg)	S.D. (%)
Copper	6.6	3	6.0	0.2	3.3
	11.7	3	12.7	1.2	9.4
	16.7	3	16.5	1.2	12.0
	21.8	3	21.4	0.6	2.8
	26.9	3	27.1	2.9	10.7
Mercury	20	3	21.1	4.0	18.9
	40	3	38.1	3.0	7.8
	60	3	60.6	6.3	10.4
	80	3	80.3	1.6	2.0
Zinc	10	3	8.8	1.3	14.7
	20	3	21.1	0.3	1.4
	30	3	30.7	1.6	5.2
	40	3	39.8	1.7	4.2
	50	3	49.5	2.5	5.1
Nickel	2	3	2.0	0.2	10.0
	5	3	5.2	0.4	7.7
	10	3	9.8	0.1	1.0
	15	3	14.8	0.8	5.4
	20	3	20.2	1.4	6.9
Cobalt	5	3	6.1	0.5	8.2
	10	3	9.7	1.1	11.3
	20	3	17.7	1.5	8.5
	30	3	30.6	2.5	8.2
	40	3	40.6	3.9	9.6
Chronium	5	5	5.7	1.3	22.8
	10	5	9.2	3.2	34.8
	20	5	20.3	8.0	39.4
	30	5	29.3	8.9	30.4
	40	5	40.5	4.2	10.4

[a] Time, 90 min; temperature, 45°C (75°C for zinc); volume, 15 ml.

Reprinted with permission from B. H. Bassos, R. F. Hirsch, and H. Letter, *Anal. Chem.*, **45**, 794 (1973). Copyright by the American Chemical Society.

into a cavity of a carbon rod electrode for subsequent flameless atomic absorption analysis.[50] Although improvements in sensitivity were not accomplished, matrix effects were significantly reduced during short electrolysis periods.

Precise and quantitative electrolysis of trace elements are difficult processes dependent on variations in stirring, pH, current density, and the presence of foreign ions. If all the metal is quantitatively deposited in the solution, the determination does not depend on the rate of deposition, but this approach can be quite time-consuming. If kinetic factors are eliminated, exhaustive electrolysis is not necessary. Exhaustive electrolysis to obtain total electrodeposition of traces is often required, however, because kinetic factors are difficult to control.

To the knowledge of the authors, few definitive investigations of the limitations of electrolytic deposition for quantitative separation of ultratrace elements have been made.[51] However very pure salts have been prepared by electrolysis in one of our laboratories.[52] Extensive investigations with radiotracers to examine thoroughly the electrodepositions of elements from very dilute solutions is a worthwhile research area for electroanalytical chemists.

B. NONELECTROLYTIC DEPOSITION

Highly sensitive preconcentration or sampling techniques for flameless atomic absorption spectroscopy have been devised. Mercury has been determined after spontaneous amalgamation on copper[53] and silver[54] wires. The deposition of cadmium, zinc, lead, and mercury on platinum coils by dipping the coils into a solution and evaporating the solvent has been reported.[55] Metal ions in solution also deposit on the surface of a tungsten alloy wire without electrolysis, presumably by an ion-exchange mechanism.[56] The tungsten wire is simply left in contact with the solution for a given period, then washed with demineralized water. The deposition of metal ions on the wire during soaking is faster than plating of the metal by electrolysis.[55]

Wires soaked in solutions of known concentration have provided analytical results up to 10^3 times more sensitive than standard flameless wire loop techniques. Detection limits for cadmium and palladium preconcentrated in this manner have been reported at 4×10^{-14} and 2×10^{-11} g, respectively.[56] An additional advantage of electrodeless procedures is that quantitative collection of the entire quantity of trace element is obviated. This approach should be well suited for quantitative spark source, electron probe, and substoichiometric isotope dilution anlaysis.

C. EVAPORATION

Evaporation in closed chambers purged with clean air continues to be the preferred concentration technique. Contamination from the heating apparatus and the evaporation vessel can be reduced and effectively controlled (see Figures 3 and 4). A Teflon chamber is used when hydrofluoric acid is a component of the evaporation mixture. These miniaturized "clean hoods" are effective in eliminating iron, aluminum, and othe components of ordinary laboratory dust. They can be easily fabricated in a glass or machine shop and should be in the possession of every trace analyst, especially those who do not have access to laminar-flow clean hoods.

Simply covering solutions during evaporation in open atmospheres is not enough to prevent airborne contamination.[8] Even in a laminar-flow hood problems were encountered in the preconcentration of ultrapure water. After 3 liters of water in a precleaned Teflon vessel was concentrated to 10 ml on a ceramic-top hot plate in a laminar-flow clean hood, analysis of the concentrate indicated high concentrations of cations ($\sim$20 ng/ml) in the original sample.

It was discovered that the airflow regulator of the clean hood had been changed to the partial exhaust mode (50% of air from the HEPA filter was exhausted through the exhaust vent). When 100% of the air was blown over the work area inside the clean hood, trace elements were detected at less than 0.1 ng/ml. This suggested that unfiltered room air was pulled directly into the work area during operation of the hood in the partial exhaust mode, even though the output of the hood, 1000 ft^3 of air per minute, was 50% larger than the exhaust capacity. By decreasing the opening to the work area by setting up a Plexiglas shield, suction of room air during partial exhaust has been minimized. It is good practice, however, to evaporate samples within gas-purged assemblies (Figure 4), when vapors must be exhausted. No contamination of aqueous solutions in open containers has been detected by mass spectrometry or neutron activation when the exhaust vent was completely closed.

Evaporation by infrared radiation concentrates solutions without boiling, thereby minimizing loss of trace residues. Sources of high energy infrared sealed in translucent quartz* provide efficient evaporation of liquids through energy absorption in the upper layer of the liquid. The balance of the liquid placed in a shallow quartz evaporating vessel remains cool (see Figure 13). The entire assembly should be placed in a laminar-flow clean hood during the evaporation. The infrared radiator consists of a fused waterproof sheath of vitreous silica, with a horizontal fused silica handle. The heating elements, enclosed completely in the sheath, are well protected

* Quartz Products Corp., 688 Somerset St., P.O. Box 628, Plainfield, N.J.

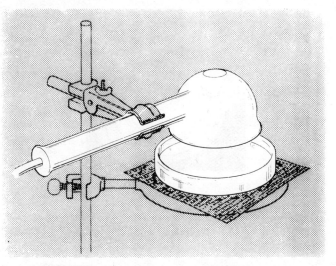

Fig. 13.   Sealed quartz infrared heater.

from corrosive vapors. The heated product is also protected from any possible contamination from the heating elements. The supporting stand and ring assembly should be constructed from Teflon or ceramic-coated material.

Effective evaporation of liquids requires shallow quartz vessels. A radiator at a distance of 4 cm above the surface of a dish filled with water can evaporate up to 0.4 l/hr. Although trace elements in bulk aqueous samples can be conveniently concentrated by evaporation in this way, highly acidic or corrosive solutions are more effectively protected from contamination by using completely closed systems as described previously.

Boutron employed nonboiling evaporation for concentrating sodium, magnesium, potassium, calcium, manganese, and iron in snow samples collected in Antarctica.[57] During this investigation, 100 ng/g standards concentrated in a laminar-flow hood and in a regular laboratory environment were analyzed. The following results, expressed in nanograms per gram, are for the laminar-flow hood and (in parenthesis) for the open atmosphere: Na, 10.2 (13.8); Mg, 10.0 (12.4); K, 10.5 (13.7); Ca, 9.1 (9.1); and Fe, 9.3 (20.6). Thus the need for clean hoods is clearly demonstrated.

The melted sample (200–500 ml) was evaporated over a 3–7 hr period to 10 ml in a 2 liter vitreous silica bulb placed in a double-walled borosilicate vessel. To prevent the sample from boiling, the temperature of the vessel was regulated by a bath of circulating silicone fluid (Figure 14). After the original volume was reduced to 10 ml, 12 ml of high-purity aqua regia was added and the resulting solution was evaporated to 1 ml. The residue was

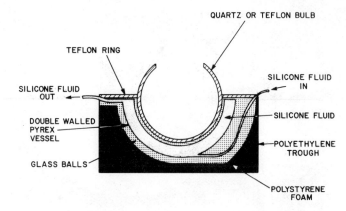

Fig. 14. Apparatus for concentration of dilute solutions by nonboiling evaporation. Reprinted with permission from C. Boutron, *Anal. Chim. Acta*, **61**, 140 (1972). Copyright Elsevier Scientific Publishing Company.

finally dissolved in 9 ml of 5% nitric acid prepared by diluting ultrapure acid with ultrapure water that had been purified by ion exchange and double distillation in vitreous silica. The water showed the following analysis (ng/g): Na, < 0.1; Mg, < 0.2; K, < 0.5; Ca, < 0.5; Mn, < 0.8; and Fe, < 1.0. No data were reported for trace metal content of the acids. However the purity of the concentrated acids used in this procedure is critical, since the acid contributes greater than 50% to the total volume of the final solution.

To reduce losses by adsorption, the evaporation chamber should be as small as possible, and enough pure acid should be initially added to make the sample 0.1 M. Alternatively, aliquots of the sample can be repeatedly evaporated to near dryness into a small calibrated vessel. The sample could then be finally diluted to a known volume and sampled. In this way losses by transfer are minimized. Investigations with radiotracers in many cases can help to identify problems encountered in the recovery of traces during concentration, and reasonable corrections for losses can be determined.

D. FREEZE–DRYING

Freeze-drying currently provides an excellent means of concentrating aqueous solutions. The principal advantage offered by this method over evaporation is the low temperature at which lyophilization occurs. Exposure of samples to prolonged heating in containers that may contribute to contamination is prevented. Since diffusion and leaching of trace elements from container walls is accelerated at elevated temperatures, contamination is significantly reduced during the freeze-drying process. The method, of course, consumes more time than evaporation at elevated temperatures.

Natural, fresh, or salt water samples are readily concentrated after being acidified to pH 1.0 with pure HNO_3 or HCl to prevent losses of trace elements by adsorption or precipitation. These are then lyophilized to remove excess water. Biological fluids and tissues are frozen directly, then dehydrated in vacuum chambers, leaving completely dry residues. Since the vapor pressures of potentially volatile traces are greater at reduced pressure, losses could occur during lyophilization. However the losses would be less than those encountered during evaporation at elevated temperatures. Detailed investigation of the retention of organomercury compounds when freeze-drying biological materials have indicated mean mercury losses of less than 3%.[58]

Commercial apparatus for freeze-drying is available, but most equipment must be modified to concentrate large volumes of aqueous samples for subsequent quantitative determinations of traces (see appendix). Most lyophilizers have been designed for drying bulk quantities of food or pharmaceutical products. The primary disadvantages of these models for analytical work include (1) bulky stainless steel cabinets or drying chambers, (2) vessels for liquids constructed from unsuitable soft glass materials, and (3) frequent presence of vacuum seals that depend on use of stopcock grease or rubber O-rings. Custom-designed units are necessary for trace analysis.

The unit appearing in Figure 15 has been used successfully for drying ultrapure materials.* The differences encountered in drying crystalline hydrates and solutions are typified by sodium carbonate. When sodium carbonate monohydrate containing 14.5% water was placed in the lyophilizer, no water was removed from the sample. When the hydrate is dissolved in water and the solution is lyophilized, however, a product containing about 0.5% water is obtained. Here the rapid freezing of the solution does not permit the formation of a crystalline hydrate that strongly retains water.

When quantitative recovery of the residue is unimportant, small aqueous samples (1–10 ml) can be dried by the following technique. Freeze the sample in the thinnest film possible on the walls of a 10 ml Teflon beaker by rotating the beaker in a dry ice acetone bath. Place the beaker in a small polypropylene vacuum desiccator and dry *in vacuo*. The aqueous sample can also be frozen on the walls of quartz tubes. The tubes are then placed in a beaker and dried in an efficient vacuum system. Such rapid, inexpensive techniques have been used to lyophilize small quantities of serum for subsequent activation analysis.[59]

* Virtis Co., Inc., Gardiner, New York.

Fig. 15.   Freeze-drying apparatus.

E.   ION–EXCHANGE AND CHELATING RESINS

The value of ion-exchange methods for separations and concentrations in analytical chemistry has been well established. The potential applicability of exchange resins for removing ultratraces from large volumes of dilute solutions is amply documented by the highly efficient demineralization of water in mixed-bed columns. Selective removal of traces from media more complicated than water has been accomplished by incorporating specific functional groups onto the resin matrix. The number of such chelating resins previously described by Schmuckler[60] has been increased by recent additions. Rohm and Haas has introduced a resin, XE-243, for concentrating 2–3 μg of boron. Chelating agents such as dithizone and 8-quinolinol have also been coupled to benzidinecarboxymethyl cellulose to recover trace elements from seawater.[61] A variety of chelating resins of 8-quinolinol have been described, and their applications for separating metals from solution have been reviewed.[62] Another specific resin for concentrating ultratraces of copper, poly-(triaminophenolglyoxal), has been synthesized[63] and made commercially available.*

In spite of the proven utility of ion and chelating exchange resins for retention of ultratraces from dilute solutions, several problems significantly impede their practical use in quantitative ultratrace analysis. First, the quantitative retention and recovery of submicrogram amounts of trace elements as cationic or anionic species remain a largely unexplored frontier. The literature lacks reliable reports that document sufficiently high recoveries of nanogram amounts of trace cations and anions, once they are retained by suitable resins. Most previous evaluations of resins have been limited to applications in which the separation, concentration, or recovery of microgram or larger amounts of impurities was investigated. In this concentration range, extremely valuable contributions in quantitative analysis have been made.

At the submicrogram level, several inherent properties of the resin may restrict their use. The present limits for purifying ion exchange resins appears to be the microgram per gram range for transition elements. Neither the synthesis of resins nor their preparation by appropriate cleaning procedures has yet been envisioned to yield products containing nanogram per gram levels of impurities. The irreversibility of the ion-exchange process for nanogram quantities of traces due to penetration into and entrapment by the ion-exchange matrix, to hydrolysis, to precipitation, or even to adsorption onto the walls of the column, are potential perils. Exchange rates and behavior of ultratraces versus pH must also be known.

Nevertheless, successful quantitative separations of submicrogram quantities of cations or anions should be possible in several general categories. This is illustrated by the recent quantitative separation of 25–125 ng of iron from a phosphate-chloride medium by anion exchange on Amberlite IR 400 resin.†[64] The $FeCl_4^-$ comples was retained by the resin, while phosphate was eluted with 5.0 M HCl. Iron was then recovered (95% or better) by elution with 0.15 M HCl-1% hydrazine. The elution of iron and phosphate was monitored by use of the isotopes ^{59}Fe and ^{32}P. Quantitative data for the separations are reported in Tables 6 and 7.

To avoid the complication of losses of traces during recovery from the resin, direct analyses of the exchange medium can be made. The development of ion-exchange membranes and resin-loaded papers facilitates direct measurements of traces on the exchange medium. Membranes, freed from

* Ionics Research Inc., Houston, Texas.
† Rohm and Haas Co., Philadelphia.

TABLE 6. Quantitative Separation of $^{59}Fe^{3+}$ (125 ng)

Column Number	Activity of 5.0 M HCl Eluate	Activity of 0.15 M HCl-1% NH_2NH_2	Retention[a] (%)	Recovery[b] (%)
2	867	445440	99.8	100
3	3457	445585	99.2	100
		445455[c]		

Quantitative Separation of $^{59}Fe^{3+}$ (50 ng)

Column Number	Activity of 5.0 M Eluate	Activity of 0.15 M HCl-1% NH_2NH_2	Activity of Resin	Retention[a] (%)	Recovery (%)
1	10506	889270	3953	98.8	98.9
2	7689	908710	2177	99.1	101
3	4508	887975	12625	99.5	98.8
		898315[c]			

[a] Percentage of initial activity retained by the resin.
[b] Percentage of original activity eluted in 0.15 M HCl-1% NH_2NH_2.
[c] Total initial activity of sample applied to the column (mean of 2 or 3 experiments).

TABLE 7. Quantitative Separation of $^{59}Fe^{3+}$ (25 ng)

Column Number	Activity of 5.0 M HCl Eluate	Activity of 0.15 M HCl-1% NH_2NH_2	Retention[a] (%)	Recovery (%)
1	7145	428910	98.4	97.3
2	14671	427770	96.6	97.1
3	13352	427910	96.9	97.1
		440697[b]		

Quantitative Separation of $^{59}Fe^{3+}$ (10 ng)

Column Number	Activity of 5.0 M HCl	Activity of Resin	Activity of 0.15 M HCl-1% NH_2NH_2	Retention[a] (%)	Recovery (%)
1	262	1049	197152	99.9	97.6
2	250	673	197277	99.9	97.7
3	4180	409	191821	97.9	95.0
			201868[b]		

[a] Percentage of initial activity retained by the resin.
[b] Total initial activity of the sample applied to the column (mean of 2 or 3 experiments).

TABLE 8. Physical Properties of Ion-Exchange Resin-Loaded Papers

	Papers			
Property	SA-2	WA-2	SB-2	WB-2
Resin	Amberlite[a] IR-120	Amberlite[a] IRC-50	Amberlite[a] IRA-400	Amberlite[a] IR-4B
Resin type	Strong acid	Weak acid	Strong base	Weak base
Percentage of resin in paper	45–50%	45–50%	45–50%	45–50%
Resin form supplied	Na^+	H^+	Cl^-	OH^-
Paper color	Tan	White	Cream	Yellow
Thickness (mils)	14	14	14	14
Wet strength	Good	Good	Good	Good
Flow rate	Fast	Fast	Fast	Fast
Approximate exchange capacity (meq./g dry resin)	4	10	3.3	10

[a] Amberlite is the registered trademark of Rohm and Haas Co., Philadelphia.

trace elements by washing with 40% sulfuric acid, were used to preconcentrate manganese, iron, and cobalt.[65] A more suitable collection of traces is provided by ion-exchange resin-loaded papers. The variety available commercially* and their physical properties are reported in Table 8. These papers have been especially useful for concentrating trace elements that could then be determined directly by X-ray fluorescence. Traces of iron, nickel, cobalt, manganese, and calcium in terephthalic acid were collected on disks of SA-2 cation exchange paper after the organic compound was ashed and dissolved by sequential acid treatments. Recoveries of 1–2 μg of these traces were, respectively, 82, 100, 91, 100, and 109%.[66] Chromium, titanium, and molybdenum were converted to the ammonium salts and collected on SB-2 anion exchange paper. Recoveries for these elements were 72, 84, and 63%. When the quantities of all elements were increased to the 2–4 μg range, the respective recoveries decreased to 73, 85, 84, 80, 94, 63, 68, and 42%. The recovery data illustrate that exchange capacities of papers must not be exceeded and that recovery values must be carefully evaluated in preconcentration procedures, since these factors impact significantly on the accuracy of a subsequently performed direct analysis.

Submicrogram quantities of nickel and vanadium originally present in 10 g samples of petroleum have also been concentrated on cation exchange paper and subsequently determined by X-ray fluorescence.[67-68] Luke[69]

* Reeve Angel, Clifton, N.J.

concentrated either 4 or 20 μg of 18 cations and 18 anionic metals from 1 ml of solution onto strongly acid and strongly basic cation and anion exchange papers. Although recoveries for cations were 87–100% and for anions 64–100%, solutions had to remain in contact with the paper for up to 3 hr.

A well-planned and carefully executed investigation of the retention of microgram amounts of cations by SA-2 paper disks and of anions by SB-2 ion-exchange loaded papers has been reported.[70] Parameters such as percentage collection as a function of pH, variation of percentage collection with the number of filtrations, the effect of solution volume on percentage retention, the effect of filtration rate, and the effect of alkali salt concentrations were investigated. Similar exhaustive studies of factors affecting the direct X-ray fluorescence determination of traces on the paper disks were also reported.

Progress in applying ion-exchange techniques in ultratrace analysis depends primarily on the analytical chemist's ability to determine with highly sensitive methods the exchange behavior of submicrogram amounts of traces. Clearly, radioisotope techniques in combination with contamination control promises to be important in extending to the submicrogram region the practical utility of ion exchange methods for accurate quantitative analysis. For example, the successful substoichiometric radioisotope dilution determination of Fe^{3+} was accomplished by isolating the anionic EDTA complex from excess Fe^{3+} by passage through a cation exchange column.[71] This is one example of the possible achievements in quantitative measurement at the submicrogram level made possible by radioisotope and ion-exchange techniques.

VII. STIRRING

Stirring is a routine, simple operation for mixing solutions or dissolving samples in the analytical laboratory. When ultratrace elements are a concern, abrasion of the container or the stirrer can be a serious problem. Conventional magnetic stirrers encased in polyethylene or Teflon, for example, are abraded on contact with breakers or flasks. Abrasion can be minimized by adopting Teflon magnetic stirrers with spinning rings.* Magnetic bars, encased in quartz, have also been fabricated by one of the authors for mixing large volumes of solutions during electrolysis.[72]

Commercial units are available in which liquids can be stirred and heated simultaneously either in round-bottomed flasks or in cylindrical bottles.†

* Cole-Parmer Instrument Co., Chicago.
† Glas-Cal Apparatus Co., Tene Haute, Ind.

The stirring devices consist of sets of electromagnetic coils that are built within the cylindrical housing of the heating mantles. Such assemblies provide even heating and stirring, eliminate bumping, and replace the conventional overhead stirring systems. The replacement of asbestos in heating mantles with a polyfluorocarbon resin fabric is particularly attractive for minimizing particulate contamination. Although magnetic stirring obviates stirring motors, adequate mixing of large volumes of solutions may require the higher efficiency of a stirrer powered by an appropriate motor assembly. In this case contamination of solutions by fallout, dust, oil, and particulates from the stirring motor assembly is eliminated by remote stirring of the liquid with a flexible shaft.* The solution is also placed in a laminar-flow hood and covered during stirring.

Nitrogen, argon, or other inert gases suitably dried and passed through 0.2 μm membrane filters can be employed for mixing 100 ml or larger amounts of solutions in Teflon containers or in 500 ml quartz bubblers. This approach is hampered by violent bubbling, which can introduce losses of the solution by the formation of a fine spray that is swept from the mixing vessel. In the dissolution of solids, rapid shaking of the mixture in Teflon or vitreous silica bottles should be substituted for stirring whenever possible. Since no stirring bar or other object is placed in direct contact with the solution, contamination by abrasion is minimized.

VIII. FILTRATION

Successful application of filtration in ultratrace work requires unusual attention to detail. Problem areas include (1) the filter paper itself, (2) ashing the filter plus sample, (3) the filter housing, and (4) airborne traces introduced by vacuum filtration. The first two problems can be reduced most effectively by a judicious selection of appropriate membrane and paper filters, and adequate cleaning before use.

The most popular filtering materials, along with the corresponding mean pore sizes, are listed in Table 9. Most materials are not inert. Paper is attacked by concentrated solutions of acids and alkalies; sintered borosilicate glass and glass filter mats are not resistant to concentrated bases. Teflon membranes resist acids and bases but are hydrophobic and must be wetted with methanol or ethanol before an aqueous solution is added. The polymeric membranes provide a complete range of chemical resistance. Compatibility tables supplied by the manufacturers (see Appendix) should be consulted for specific applications.

* Curtin Scientific Co., Rockville, Md.

TABLE 9. Filtering Media

Material	Mean Pore Size (μm)
Paper (cellulose)	
Coarse	6
Fine	2
Sintered glass	
Coarse	40–60
Medium	10–15
Fine	4–5.5
Ultrafine	0.9–1.4
Glass fiber mats	0.3–8.0
Silver	0.2–5.0
Linear polyethylene	
Medium coarse	60–70
Medium	30–40
Membranes (polymeric)	
Cellulose	0.2–0.45
Cellulose acetate	0.2–5.0
Cellulose esters (mixed)	0.025–8.0
Polypropylene	0.9–10
Polycarbonate	0.1–8.0
Nylon	0.45–3.0
Polyvinyl chloride	0.6–2.0
Teflon (TFE)	0.2–10.0

Since one of the most popular filter materials is paper, virtually 100% cellulose, the adsorption properties of cellulose must be taken into consideration during filtering of pure or dilute solutions. Cellulose is widely used as an adsorbant in thin-layer chromatography (TLC) for the separation of inorganic ions, organics, and biochemicals. Filtration through paper, therefore, can parallel the selective adsorption observed in TLC. Of course the flow rate during filtration is too rapid for optimum adsorption, but the behavior of ultratraces is unpredictable.

Even though paper consists of reasonably pure cellulose, it is routinely contaminated with trace metals during manufacturing. A comprehensive tabulation of trace elements in commercially available filters has been prepared by Robertson.[1] Data for some of the commonly used filters are shown in Table 10. Filters fabricated from cellulose acetate, polypropylene, fluorocarbons, and polycarbonate, are available.* In view of trace elemental contents given in Table 10, it is clear that most of these filters must be precleaned before every use.

* Millipore Corp., Bedford, Mass.; General Electric Co., Schenectady, N.Y.; Gelman Instrument Co., Ann Arbor, Mich., Nuclepore Corp., Pleasanton, Calif.

TABLE 10. Trace Elements in Filter Papers ($\mu g/g$)

Filter	Cu	Br	Na	Al	S	Cl	K	Sc	V	Cr	Mn	Fe	Co	Zn	Sb
											Elements				
Mitex (Teflon)	2	<0.5	3	8	—	3	—	—	<0.03	—	0.3	—	—	—	—
Whatman 541	<1	<1	13	1	—	60	—	—	<0.03	—	0.3	—	—	—	—
Whatman Chromatograph #3	—	—	0.9	1	—	6	—	—	1	—	0.2	—	1	—	—
Millipore HA	—	—	—	—	—	8×10^{-4}	17.6	—	0.33	0.013	2.4	0.039	—	—	—

Source: Ref. 1.

Soaking filters in $CHCl_3$ solutions of chelating agents such as pyrrolidine-dithiocarbamic acid or dithizone should be effective for removal of the common di- and trivalent cations. Presoaking in acid solutions, 0.1 M oxalic acid or citric acid, or in basic EDTA solutions followed by rinsing with distilled water, is highly recommended.

Many polymeric membranes contain detergents. However detergent-free membranes are available from suppliers upon request. Detergents can contribute trace metals and can inhibit the growth of bacteria in microbiological applications.

Contamination by the filter housing is minimized by using all-plastic filtration assemblies. Care must be exercised to prevent contact of solutions with plastics containing inorganic pigments. Plastic filter housings that are green, orange, or blue should not be used.

Airborne contamination via vacuum suction is avoided by performing the filtration in a clean hood. An alternative procedure, pressure filtration (Figure 16), is most desirable for removing insoluble particles and for preserving the quality of the filtrate. The solution to be filtered is transferred to Teflon bottles (1) and forced by 30 psi of filtered argon pressure to pass through filter (2), which contains a precleaned 0.2 μm porosity membrane filter. The filtered solution is then collected and stored in a second Teflon (TFE) container. By connecting the filtered argon supply to a special outlet on the storage vessel (2), the filtrate can be dispensed by pressure without opening the container to the atmosphere. Systems of this type are well suited for separating particulates from natural water samples or for ultrapurification of reagents (Chapter 5).

Frits of porous glass, vitreous silica, and polyethylene are recent innovations that have expanded significantly the utility of ultrapurity filtration techniques. The major advantage is provided by the suitability of these media for treating acids and reactive chemicals. Prolonged leaching of fused vitreous silica frits in HCl and HNO_3 followed by rinsing in pure water is necessary to remove iron, a significant contaminant.

Environmental analyses of natural waters often require filtration. Suspended solids must be separated before determination of dissolved cations. The filter medium selected to retain the particles can directly affect the analytical results through contamination, as well as through its efficacy for the collection, which is usually a function of pore size. In a systematic investigation of the effects of the filter on analytical results, it was established that iron and aluminum concentrations in river water decreased with smaller filter pore size.[65] Silicon and magnesium levels were constant. In these studies, silver metal (Flotronics, pore sizes 5–0.2 μm) and cellulose ester (Millipore, 5–0.025 μm), filters exhibited different retention behaviors when identical filtration methods were used and when nominal pore sizes were

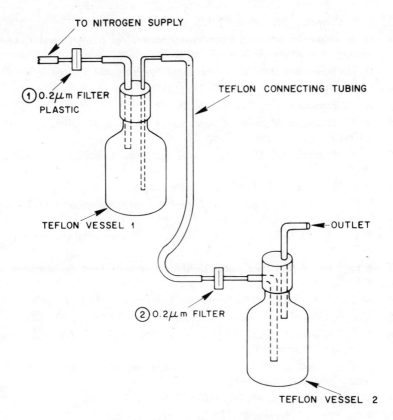

Fig. 16. Pressure filtration system.

equivalent. Although the metal filter was more effective in removing particulates than the Millipore filter, use of the latter in ultratrace cation analysis is preferred, to minimize contamination problems.

The importance of filtration in trace analysis and in ultrapurification has increased significantly in recent years. Filters for collecting airborne particulates, for purifying gases, for applications in neutron activation analysis, and for quantitative recovery of micro amounts of precipitates, are but a few of the important uses.

REFERENCES

1. D. E. Robertson, in *Ultrapurity Methods and Techniques,* M. Zief and R. Speights, eds., Dekker, New York, 1972, pp. 207–253.
2. D. E. Robertson, *Anal. Chem.,* **40,** 1067 (1968).

3. L. H. Cooper, *J. Mar. Res.,* **17,** 128–132 (1958).

4. D. N. Hume, in *Advances in Chemistry* Series, Vol. 67, American Chemical Society, Washington, D.C., 1967, pp. 30–41.

5. G. Thompson and D. C. Bankstron, *Appl. Spectrosc.* **24** (2), 210 (1970).

6. L. Ducret, *Anal. Chim. Acta,* **17,** 213 (1957).

7. X. I. Zilbersthein et al., *Zavod. Lab.,* **28,** 680 (1962).

8. R. E. Thiers, in *Methods of Biochemical Analysis,* D. Glick, ed., Vol. 5, Interscience, New York, 1957, pp. 274–309.

9. J. W. Mitchell and J. E. Riley, Bell Telephone Laboratories Publication, 1972.

10. D. L. Malm and J. W. Mitchell, unpublished procedures.

11. I. P. Alimarin, ed., *Analysis of High-Purity Materials,* Israel Program for Scientific Translations, Jerusalem, 1968, pp. 17–21.

12. P. F. Schmidt, Bell Laboratories, Allentown, Pa., personal communications, 1973.

13. J. Labrador, J. M. Laviec, and J. Lorthioir-Pommier, *Prod. Probl. Pharm.,* **26,** 622–636 (1971).

14. M. Zief, J. Horvath, and N. Theodorou, Lab Practice **25**(4), 215 (1976).

15. J. S. Nesbitt and E. M. Knapp, U.S. Patent 3,577,322, May 1971.

16. A. D. Pearson, and W. G. French, *Record,* **50,** 106 (April 1972), Bell Laboratories, Inc., Murray Hill, N.J.

17. J. W. Mitchell and W. R. Northover, paper presented at *Am. Ceram. Soc.,* Washington, D.C., June 1972.

18. A. J. Rastron, *Sci. J.,* **69,** July 1967.

19. W. A. Perfer, *J. Metals,* **48,** 487, May 1965.

20. A. E. Jenkins, B. Harris, and L. Baker, *Symposium on Metal at High Pressure and High Temperature,* Metallurgical Society, AIME Conference, Vol. 22, Gordon & Breach, New York, 1964, pp. 23–43.

21. B. Harris, E. G. Price, and A. E. Jenkins, *Symposium on Peaceful Uses of Atom Energy,* Melbourne, University, Press, Melbourne, Australia, 1958, pp. 221–225.

22. W. Grunwald, Dissertation, Stuttgart, 1973.

23. Analytical Methods Committee, *Analyst* (*London*), **79,** 397 (1954).

24. J. I. Hoffman and G. E. F. Lundell, *J. Res. Nat. Bur. Stand.,* **22,** 465 (1939).

25. R. E. Thiers, in *Trace Analysis,* J. H. Yoe and H. J. Koch, Jr., eds., Wiley, New York, 1957, pp. 637–666.

26. C. A. Evans, Jr., and G. H. Morrison, *Anal. Chem.,* **40,** 869 (1968).

27. C. E. Gleit and W. D. Holland, *Anal. Chem.,* **34,** 1455 (1962).

28. J. W. Mair, Jr., and H. G. Day, *Anal. Chem.,* **44,** 2015 (1972).

29. B. Bernas, *Anal. Chem.,* **40,** 1682 (1968).

30. B. Bernas, *Am. Lab.,* p. 41, August 1973.

31. J. P. Riley and H. P. Williams, *Mikrochim. Acta*, 516 (1959).

32. J. Ito, *Bull. Chem. Soc. Jap.*, **35**, 225 (1962).

33. F. J. Langmyhr and S. Sveen, *Anal. Chim. Acta*, **32**, 1 (1965).

34. F. J. Langmyhr and P. E. Paus, *Anal. Chim. Acta*, **49**, 358 (1970).

35. W. Holak, B. Krinitz, and J. C. Williams, *J. AOAC*, **55**, 741 (1972).

36. J. W. Mitchell and D. L. Nash, *Anal. Chem.*, **46**, 326, (1974).

37. A. F. Voigt, personal communication.

38. A. Mizuike, in *Trace Analysis, Physical Methods*, G. H. Morrison, ed., Wiley, New York, 1965.

39. J. Inezdy, *Period. Polytech.*, **14**, 149 (1970).

40. P. Jannasch and R. Seiske, *J. Chem. Soc.*, **114**, 460 (1918).

41. P. Jannasch and O. Laubi, *J. Prakt. Chem.*, **97**, 150 (1918).

42. K. G. Isave and L. G. Zhuralev, *Anal. Abstr.*, **7**, 13183 (1960).

43. G. Tölg, *Talanta*, **19**, 1489 (1972).

44. M. Haizzinsky, *J. Chim. Phys.*, **43**, 21 (1946).

45. L. B. Rogers, *Anal. Chem.*, **22**, 1386 (1950).

46. J. C. Griess, Jr., and L. B. Rogers, *J. Electrochem. Soc.*, **95**, 129 (1949).

47. B. H. Bassos, R. F. Hirsch, and A. G. Puchuto, *Anal. Chem.*, **43**, 1503 (1971).

48. B. H. Bassos, F. J. Berlandi, T. E. Neal, and H. B. Mark, Jr., *Anal. Chem.*, **37**, 1653 (1965).

49. B. H. Bassos, R. F. Hirsch, and H. Letterman, *Anal. Chem.*, **45**, 792 (1973).

50. C. Fairless and A. J. Bard, *Anal. Lett.*, **5**, 433 (1972).

51. G. Tölg, *Vom Wasser*, Vol. 40, Verlag-Chemie, Weinheim/Bergstr., 1973, pp. 181–206.

52. M. Zief and J. Horvath, *Lab. Pract.*, **23**(4), 175 (1974).

53. H. Brandenberger and H. Bader, *Helv. Chim. Acta*, **50**, 1409 (1967).

54. M. J. Fishman, *Anal. Chem.*, **42**, 1462 (1970).

55. M. Soicheio, M. Makota, and M. Hirokozu, *Bunseki Kagaku*, **20**, 1177 (1971).

56. M. P. Newton, J. V. Chauvin, and P. G. Davis, *Anal. Lett.*, **6** (1), 89 (1973).

57. C. Boutron, *Anal. Chim. Acta*, **61**, 140 (1972).

58. P. E. LaFleur, *Anal. Chem.*, **45**, 1534 (1973).

59. J. W. Young and G. D. Christian, *Anal. Chem.*, **45**, 1296 (1973).

60. G. Schmuckler, *Talanta*, **12**, 281–290 (1965).

61. A. J. Bauman, H. H. Weetall, and N. Weliky, *Anal. Chem.*, **39**, 932 (1967).

62. F. Vernon and H. Eccles, *Anal. Chim. Acta*, **63**, 403 (1973).

63. A. Zlatkis, W. Bruening, and E. Bayer, *Anal. Chem.*, **42**, 1201 (1970).

64. J. W. Mitchell and V. Gibbs, *Anal. Chem.*, in press.

65. W. T. Grubb and P. D. Zemany, *Nature,* **176,** 221 (1955).
66. R. Wagemann and G. J. Brunskill, *Int. J. Environ. Anal. Chem.,* **4,** 75 (1975).
67. J. G. Bergmann, C. H. Ehrhardt, L. Granatelli, and J. L. Janik, *Anal. Chem.,* **39,** 1331 (1967).
68. J. G. Bergmann, C. H. Ehrhardt, L. Granatelli, and J. L. Janik, *Anal. Chem.,* **39,** 1258 (1967).
69. C. L. Luke, *Anal. Chem.,* **36,** 319 (1964).
70. W. J. Campbell, E. F. Spano, and T. E. Green, *Anal. Chem.,* **38,** 987 (1966).
71. J. Starý and J. Ružička, *Talanta,* **8,** 775 (1968).
72. M. Zief, unpublished work.

SELECTED METHODS FOR DETERMINING ULTRATRACE ELEMENTS IN REAGENTS AND MATERIALS

Many monographs, reviews, and manuscripts have discussed various trace methods of analysis comprehensively. The general principles and theory of the methods, advances in procedures and in instrumentation, and applications have been especially well treated. Consequently these topics are not covered in detail in this chapter, which describes the problems and difficulties associated with several trace methods that are particularly useful for characterizing ultrapure materials. Factors affecting the acccuracy of the method, advantages and limitations, and applications for which the technique may be uniquely suited are highlighted. Instrumental methods that can be combined with careful chemical procedures to produce high accuracy and precision in determining submicrogram quantities of traces are emphasized.

The omission of a method does not imply a serious limitation of the technique for ultratrace work. We simply stress methods used by the authors and their colleagues or techniques with sufficient literature references to demonstrate broad applicability for quantitatively determining ultratraces in reagents or materials. The text contains comprehensive references for the various methods.

I. NEUTRON ACTIVATION ANALYSIS (NAA)

A. GENERAL

1. Advantages

Since the work of G. Hevesy and H. Levi[1] in 1935, activation analysis has become one of the most important trace methods for elemental analysis. Because of its extreme sensitivity, many applications have been made in high purity materials, in medical and biological fields, in criminology, in archaeology, in air and water pollution control, and in nutrition.[2] The broad applicability of the method is demonstrated by the list of elements (Table 1) that are determinable by thermal neutron activation. The approximate detection limits given are those obtained by Jenkins and Smales in the absence of interferences.[3] These investigators assumed that the samples

TABLE 1.   Thermal Neutron Activation Sensitivities[a]

Limit of Detection (g)	Element
10^{-12}	Eu, Dy
10^{-11}	Mn, Pd, In, Sm, Ho, Re, Ir, Au
10^{-10}	Na, Sc, Cu, Ga, As, Br, Kr, Y, Sb, Pr, Tb, La, Er, Yb, Ta, W, Th, U
10^{-9}	P, Ar, K, Rb, Co, Ru, Cd, Cs, Ba, Ce, Nd, Gd, Hf, Os, Pt, Hg
10^{-8}	Cl, Si, Ni, Zn, Ge, Se, Mo, Ag, Sn, I, Xe, Tl
10^{-7}	Ca, Sr, Fe, Zr, Bi
10^{-5}	Mg, Pb

[a] Sensitivities for n,γ reactions at a neutron flux 10^{12} neutrons/$(cm^2)(sec)$; according to Jenkins and Smales.[3]

were activated by a neutron flux of 10^{12} n/(cm^2) (sec) up to saturation, or a maximum of one month for long-lived radionuclides. The detection limit was assumed to be the amount of element possessing an activity of ≥ 100 disintegrations per minute 2 hr after the irradiation. Sensitivities for elements with isotopes having half-lives ≤ 10 min. are listed in Table 2. It is apparent from these tables that more than two-thirds of the elements

TABLE 2.   Sensitivity of Neutron Activation for the Determination of Short-Lived Radioisotopes

Element	Activation Reaction	Half-life		Sensitivity[a] (g)
O	$^{16}O(n, p)^{16}N$	7.4	sec	—
F	$^{19}F(n, p)^{19}O$	24.9	sec	10^{-7}
N	$^{14}N(n, 2n)^{13}N$	10	min	—
Al	$^{27}Al(n, \gamma)^{28}Al$	2.3	min	10^{-9}
Si	$^{28}Si(n, p)^{28}Al$	2.3	min	10^{-9}
Ca	$^{48}Ca(n, \gamma)^{49}Ca$	8.8	min	10^{-8}
Sc	$^{45}Sc(n, \gamma)^{46m}Sc$	19.5	sec	10^{-9}
Ti	$^{50}Ti(n, \gamma)^{51}Ti$	5.8	min	10^{-8}
V	$^{51}V(n, \gamma)^{52}V$	3.76	min	10^{-10}
Cr	$^{52}Cr(n, p)^{52}V$	3.76	min	10^{-9}
Cu	$^{65}Cu(n, \gamma)^{66}Cu$	5.2	min	10^{-10}
Co	$^{59}Co(n, \gamma)^{60m}Co$	10.5	min	10^{-11}
Se	$^{76}Se(n, \gamma)^{77m}Se$	17.5	sec	—
Rb	$^{85}Rb(n, \gamma)^{86m}Rb$	1	min	10^{-9}
Nb	$^{93}Nb(n, \gamma)^{94m}Nb$	6.6	min	—
Tl	$^{205}Tl(n, \gamma)^{206}Tl$	4.2	min	10^{-8}

[a] From Ref. 7.

can be determined directly in submicrogram quantities by (n, γ) reactions, assuming a flux of 10^{12} n/(cm²)(sec). Many reactors operate routinely at fluxes of 10^{13} and some at 10^{14} n/(cm²)(sec), indicating that most elements can be analyzed within the range of 10^{-9}–10^{-14} g, a sensitivity not routinely attainable on a broad scale with other methods.

Theoretical or interference-free sensitivities can differ substantially from the practical sensitivities obtained during a given determination. Under practical conditions a radionuclide formed during activation can interfere because of photopeak overlap or Compton continuum. Sensitivity is also a function of neutron flux, detector efficiency, and measurement geometry.

In addition to extreme sensitivity, which obviates preconcentration, there is another unique advantage of activation analysis because no errors are introduced by a blank value resulting from reagent or airborne contamination during postirradiation chemical procedures. After the sample is activated, the analyst adds a quantity of carrier (usually 1–10 mg) that is orders of magnitude greater than either the amount of element in the activated sample (usually < 1.0 μg) or the quantity introduced by all contamination during chemical processing. The blank value is made insignificant and does not affect the quantitative result.

The additional advantage, working with a macro quantity of the element while determining ultratrace quantities, eliminates the difficulties associated with quantitatively reacting, separating, and recovering traces of elements. The yield of carrier recovered after chemical processing is determined and corrections are made for incomplete recovery of the element to be determined. No other analytical method provides this unique combination of advantages for the measurement of submicrogram quantities of elements.

2. Limitations

Several problems accompany analysis by neutron activation. Elements with long half-lives (> 1 year) frequently must be activated for 100 hr or longer to achieve nanogram per gram sensitivity. During this period highly radioactive samples may be produced because of matrix elements or major components. Cooling periods of several days or even weeks may be necessary to allow for decay of an undesirable isotope. Although computer programs have been developed to optimize the time of irradiation and the cooling period,[4] the turnaround time for some determinations is on the order of several weeks. Fortunately a large number of samples usually can be irradiated simultaneously. A substantial number of elements can also be determined within a few minutes or several hours by using short-lived isotopes (Table 2).

Nondestructive analysis with high resolution lithium-drifted detectors and computers has been established as an extremely valuable instrumental technique. Many determinations, however, require radiochemical group separations or selective isolation of the element of interest from an intensely emitting matrix component. Thus considerable use is made of separations methods, especially ion exchange and solvent extraction. Occasionally precipitation methods are essential. These procedures are generally time-consuming, but several operations for performing radiochemical separations can be successfully automated.[5, 6]

Activation analyses are expensive in comparison with other techniques. The superiority of this approach with respect to applicability for many ultratrace determinations has largely offset this disadvantage, however, especially in the case of semiconductor materials. Of course the most serious problem is the need for a nuclear reactor, a less versatile neutron generator, or a cyclotron.

3.   Sample Preparation

a.   Contamination Problems

Substantial freedom from blank problems makes activation analysis a preferred technique. Although blanks resulting from postirradiation handling and processing of the sample do not affect the analytical result, handling of the sample before activation must be performed in a way that causes no contamination with the elements to be determined or with those that could be activated. This is largely accomplished by avoiding preirradiation treatment of the sample. To weigh and dry samples before packaging in appropriate irradiation containers, one should use a laminar-flow hood in a clean room.

The sample must be packaged to prevent contamination during transport to the reactor site and also during the irradiation. Samples can be contaminated with radioactive traces from the container following the irradiation. Often, for example, powdered samples can be recovered quantitatively only by breaking the quartz ampoule over a vessel containing an acid or an oxidizing agent. The sample is then dissolved by heating it along with fragments of the ampoule. Surface impurities or activated components of the quartz can be leached. Although the sample is quantitatively recovered, this radioactive contamination is interpreted as impurities in the sample, and large errors are introduced. Additionally, if an insoluble powdered sample is to be counted in the solid state, any small fragments of quartz produced by fracturing the ·container cannot be separated. If these fragments contain major or minor elements that are being determined in the sample, quantitative results will be grossly inaccurate.

b.   Liquids

Aqueous samples to be irradiated for 2 hr or less should be sealed in appropriately cleaned polyethylene tubing or vials. One of the authors has found the following procedure to be convenient. Polyethylene tubing (6.3 mm o.d.) is cut into 15.2 cm lengths with Teflon-coated pruning shears, rinsed with distilled water to remove any visible dust and particulates, washed with methanol, then with water, and leached subsequently with 1:1 HNO$_3$ (ULTREX®, J. T. Baker) for 4–6 hr at 70°C. The tubes are then rinsed sequentially in distilled H$_2$O and demineralized, doubly distilled H$_2$O. During the rinse and leaching processes the tubes are not permitted to dry. The leached tubes are finally dried on a plastic rack under an infrared lamp in the laminar-flow hood (Figure 2, Chapter 6).

The end or center of the dry tube is rotated in a gas-oxygen flame until translucent, then sealed by pressing the soft center with a pair of platinum-tipped or aluminum tongs. The tube is rinsed with the aqueous sample by dispensing from a polystyrene throwaway pipet. The sample tube containing about 3–6 ml is sealed by placing the end of the tube either in a flame or in an inverted crucible furnace, which has been provided with a quartz liner. The soft end is then pressed to provide a good seal, as previously described. The tubes are marked for identification by heating the end of a small screw driver and inscribing numbers.

Liquid samples to be irradiated for 3 hr or more at the reactor core must be packaged in precleaned, synthetic, high purity vitreous silica tubing. This tubing, cleaned as described previously for polyethylene, is more complicated to seal than polyethylene and becomes more active following irradiation. Although high purity synthetic silica is reasonably deficient in trace metals, cooling periods of several days may be necessary for decay of ^{24}Na, ^{38}Cl, ^{31}Si, and ^{125}Sb, before samples can be handled without shielding. Sealing and opening quartz tubes is facilitated by first freezing the liquid sample by rotating the tube in a beaker of liquid nitrogen.

c.   Powdered Samples

Powdered samples are packaged in either polyethylene or vitreous silica as described previously. Solid bulk samples can be conveniently packaged in aluminum containers of various shapes or wrapped in suitable aluminum foil (thickness $\geq$ 0.01 cm). In spite of careful etching of the sample after activation, complete removal of surface impurities that have diffused from the aluminum wrapping is difficult. Thin aluminum foils ($\leq$ 0.003 cm) or any aluminum container should be avoided when activating hydrated, powdered, or caustic compounds.

d. Sample Recovery

Known quantities of the irradiated samples must be taken for radiochemical processing or direct counting. Irradiated liquid samples are cooled to room temperature and known aliquots (1 ml or more) are pipetted with disposable polystyrene pipets. Irradiated powdered samples can be weighed into polyethylene beakers on a remote digital balance provided with lead shielding. Bulk samples have been weighed similarly following irradiation, etching with a suitable acid, and drying.

e. Evaluation of Procedures

The effectiveness of procedures for cleaning polyethylene vials and the extent of leaching from and adsorption on container walls must be determined to assess the effects of these variables on the quantitative analysis. The authors have used neutron activation to examine the efficiency of procedures for cleaning sample tubes. The cleaned plastic or vitreous silica tubes were dried, heat sealed, and irradiated for 20 min in a flux of thermal neutrons at 10^{13} n/(cm²)(sec). Following irradiation, the exterior of each capsule was rinsed with dilute nitric acid and dried. Opening the tube was facilitated by a scratch made near one end with a triangular file before irradiation. After irradiation the tube was inserted into a horizontal copper pipe of a slightly larger diameter until only the portion beyond the scratch was visible. The shorter section of the silica tube was carefully snapped off by inserting the free end into a second copper pipe and applying pressure. The tube was then filled with a warm 1:1 nitric acid solution to remove any trace elements deposited on the interior surface during the cleaning process. The rinse solution was transferred to a counting vial and the γ-ray spectrum was recorded. The tube was rinsed several times with water and filled with hot nitric acid. Contaminants that were capable of easily diffusing or migrating from the walls of the container were removed by leaching the capsule for 30 min at 70°C.

The γ-ray spectra of the rinse and leaching solutions and of the container were recorded with a high resolution lithium-drifted germanium detector. To obtain quantitative information, standard solutions were simultaneously irradiated with the cleaned tubes. The γ-ray energies, background corrections, and net photopeak areas were calculated from spectral data on a Honeywell 6000 computer.

In polyethylene tubing, traces of copper (2.6×10^{-8} g), manganese (3.9×10^{-9} g), sodium (1.8×10^{-7} g), and chlorine (1.1×10^{-5} g) were found in a 1.273 g sample. Analysis of the solution resulting from rinsing and from leaching the interior surface of a 10 cm irradiated polyethylene tube (4.740

g, 6.3 mm i.d., 9.5 mm o.d.) with 10 ml of 1:1 nitric acid showed no indication of these traces.

Tubes of Spectrosil and Vitreosil (6.3 mm i.d., 9.5 mm o.d. $\times$ 15 cm
long) were cleaned, sealed, irradiated, and treated in the manner previously
described. Impurities with long-lived isotopes were detected by recording
spectra after 72–100 hr of irradiation. Antimony was found to be a significant impurity in Spectrosil and was slowly removed by leaching with 1:1
HNO_3. No significant contamination by chromium, cobalt, iron, zinc, or
other transition elements with long half-lives ($T_{1/2} \geq 4$ d) was indicated.
Similar results were obtained for Vitreosil, which was highly contaminated
with antimony: 0.8% of the total antimony in this material ([Sb] > 5.0 μg/
g) was removed by leaching with HNO_3 after the irradiation.

Short-lived impurities in Spectrosil tubing with photopeaks below 0.4
MeV were masked by the β continuum from ^{31}Si. Quantitative analysis
showed 4.6 $\times$ 10^{-6} g of chlorine in 0.785 g of Spectrosil. The rinse solution
contained traces of manganese (6.16 ng), chlorine (15 ng), sodium (4.4 ng),
and copper ($<$ 1.0 ng, photopeak of ^{64}Cu visible but not statistically
detectable above background). The amounts of copper (0.84 ng), manganese
(0.05 ng), and sodium (2.8 ng) detected in the leach solution, demonstrated
the expected low level of contamination from Spectrosil. After a 15 min
irradiation, Vitreosil tubes were too radioactive to permit counting for
short-lived isotopes. Photopeaks present in the spectrum indicated that the
decay of ^{122}Sb and ^{24}Na was primarily responsible for these intense radiations. Thus no significant contamination by residual impurities remaining
on the interior surface of the tubes after cleaning was detected. Contamination resulting from leaching of the polyethylene container during irradiation
was also minimal. Only leaching of antimony from vitreous silica was significant.

The extent of loss of traces by adsorption was determined by irradiating
a series of dilute standard solutions during a 10 hr period. Table 3 gives the
γ-ray activities induced in the solutions. The average deviations of the mean
of these values, 61265 $\pm$843, 1627 $\pm$97, 1228 $\pm$76, and 5527 $\pm$184, show
the conditions for irradiation to be reproducible within $\pm$3–6%. By using
flux monitors the precision of irradiations can be improved further. No significant loss of trace elements to the walls of the containers nor any detectable contamination from the containers is indicated by these results.

f. Volatilization Losses

Losses of traces from samples by volatilization during irradiation must be
considered because of the high temperature in the reactor. Mercury is

TABLE 3.   Decay-Corrected Net Photopeak
Areas of Isotopes in Standard Solutions

$^{56}Mn^a$	$^{38}Cl^b$	$^{24}Na^c$	$^{64}Cu^d$
60177	1713	1390	5651
60121	1519	1138	5303
61817	1790	1224	5383
60966	1663	1293	5772
61368	1520	1133	20257^e
63140	1556	1190	19401^e

[a] Energy of γ, 0.847 MeV; volume of solution, 1.0 ml;
concentration, 1.06×10^{-7} g/ml.
[b] Energy of γ, 2.17 MeV; volume of solution, 1.0 ml;
concentration, 4.42×10^{-7} g/ml.
[c] Energy of γ, 2.75 MeV; volume of solution, 2.0 ml;
concentration, 2.86×10^{-7} g/ml.
[d] Energy of γ, 0.511 MeV; volume of solution, 1.0 ml;
concentration, 1.08×10^{-7} g/ml.
[e] Measured with a sodium iodide detector.

known to volatilize within the temperature range 55–140°C from samples as
an organometallic compound and also as an inorganic salt.[7] Conversely,
iodine in organic polymers was found to be completely retained in the
sample during activation.[8] Careful examination of the possibility of
volatilization must be considered if the element of interest is contained in
the sample as a volatile halide, hydride, oxide, or element (Table 4). The
only effective method for coping with this problem is considered to be the
perfect sealing of the container and measurement of the activity of the
sample together with the container by γ-spectrometry.[7] Obviously the

TABLE 4.   Elements and
Compounds Capable of Volatilizing
During Neutron Activation

$AsCl_3$	$SnBr_2$
$SbBr_3$	$Se(Br)_4$
$SbCl_3$	$Hg(Br)_2$
$GeCl_4$	PH_3
$ZrCl_4$	OsO_4
WF_6	RuO_4
WO_2Cl_2	Cl_2
Hg	Br_2
CrO_2Cl_2	I_2

container must be impermeable to the volatile components and must also be free from the element of interest.

B. PRINCIPLE OF MEASUREMENT

In activation analysis the standard containing an accurately known quantity of the element of interest is irradiated simultaneously with the sample. Provided identical irradiation conditions are met for sample and standard, the ratio of the induced activity is used to obtain the quantitative result according to

$$W_s = W_{std} \left(\frac{A_s}{A_{std}} \right) \tag{1}$$

where W_s and W_{std} are the unknown weights of the element in the sample and the known quantity in the standard, and A_s and A_{std} are the corresponding radioactivities of the element. When radiochemical separations are performed, corrections for the yield are made according to

$$W_s = W_{std} \left(\frac{Y_{std}}{Y_s} \right) \left(\frac{A_s}{A_{std}} \right) \tag{2}$$

Here Y_s and Y_{std} are the radiochemical yields of carrier recovered from the sample and the standard, respectively.

Techniques of substoichiometric separations have been introduced to eliminate the need to perform yield measurements in activation analysis. With this method the yield for the sample and standard can be made identical. Equation 2 then reduces to

$$W_s = W_{std} \frac{A_s'}{A_{std}'} \tag{3}$$

Where A_s' and A_{std}' are the activities of the substoichiometrically separated fractions. Available methods for obtaining equal yields during radiochemical separations have been reviewed,[9-11] and new methods have been recently introduced by one of the authors.[12-15].

Another advantage obtained by substoichiometrically extracting the desired element is that quantitative results are essentially free from all interferences from other traces, as well as major cationic or anionic constituents. Since the concentrations of the desired activated element (1 ng/g–1 μg/g) and of any radioactive trace element in the sample or in the standard are made negligible in comparison to the amount of carrier (1–10 mg added following activation), consumption of the substoichiometric reagent by other traces can be neglected. The matrices of the sample and of the standard solutions are made identical by adding to the synthetic standard a portion of the original nonactivated sample that weighs exactly the same as the activated sample undergoing analysis. Both the sample and standard mixtures are then treated chemically in the same way to dissolve the sample, to adjust the aqueous phase to the appropriate conditions, and to perform the substoichiometric separation. Any minor enhancement or depression of the distribution ratio of the carrier element due to major cations or anions in the aqueous phase now equally affects the extraction of the carrier from the standard and from the sample. The quantitative results calculated from the activity of the organic phases are thus independent of any chemical factors that can affect the extraction. The only interferences that now must be circumvented are due to a major component of the sample that is extracted more readily than the carrier ion, or to a major anion or complexant that depresses the distribution ratio sufficiently to yield too little activity in the organic phase. In the former case a preliminary separation of the interfering element or of the carrier is necessary before the substoichiometric extraction is performed.

Statistical errors in measuring A_s and A_{std} are related to the size of these quantities. The effect on the quantitative result W_s is minimized by closely matching (within a factor of 10 or better) the mass (W_{std}) with the anticipated mass of element in the sample (W_s). Although little effort is needed to determine Y_s and Y_{std} within $\pm 2\%$ accuracy or better, precautions must be taken to ensure that the carrier is introduced quantitatively in the same chemical form as the element in the sample, or that both are converted completely into identical chemical species during dissolution of the sample. Uncertainty in the known mass of element in the standard W_{std} is directly transformed into an error of the same magnitude in the determined value W_s.

C. STANDARDS

1. General

Synthetic comparison standards must be prepared for many neutron activation analyses, since certified reference materials similar to the sample rarely exist. Suitable standards are sometimes obtained by weighing fractional milligram amounts of zone-refined or electrolytically produced wires, pellets, or foils on a micro balance, and by irradiating, dissolving, and taking appropriate aliquots for comparison. Handling a single metallic sample facilitates quantitative recovery after irradiation. However, when milligram amounts of some metals are irradiated, excessive activity is produced. Neutron self-absorption in the standard is also possible.

2. Solids

Powdered oxides, or salts such as carbonates, acetates, or nitrates, are satisfactory for cations of interest. Ammonium salts should be used when the element is a constituent of an anion. In this laboratory the use of small quantities (mg or less) of powdered compounds was abandoned because of the low precision of quantitatively recovering the powder from its container.

Dilution of milligram quantities of the standard compound in the solid state with compounds absorbing very few neutrons (e.g., aluminum oxide, sucrose, and silica) has decreased the self-absorption of neutrons. Corrections for blanks due to traces in the solid diluent are then necessary.

3. Solutions

In this laboratory aqueous solutions have proved to be the most effective synthetic standards. Solutions containing ≥ 0.1 $\mu g/ml$ of the element of interest have been prepared by dissolving accurately weighed amounts of highly pure metals or wires in high purity nitric acid and diluting to a final volume in a plastic volumetric flask with double-distilled, deionized water. The final pH of the diluted standard is maintained at ≤ 1 to minimize adsorption of the trace element. All the precautions mentioned in Chapter 2 are taken during the preparation of the standard solution. A 2 ml portion of the resulting solution can be easily irradiated, and 0.5–1.0 ml can be accurately measured following activation.

A popular approach involves evaporating accurately measured small volumes of solutions on filter paper, polyethylene sheet, or inside a silica tube. It is important here to take into account impurities, especially sodium and chlorine, in the material on which the solution is evaporated. After activation, quantitative recovery of an evaporated residue may be difficult. For example, 100 μg quantities of cobalt deposited by evaporation in quartz tubes at this laboratory could not be recovered with a precision better than $\pm 20\%$, while a series of solutions irradiated sequentially had induced ^{60}Co activities within $\pm 1\%$. If filter paper is sufficiently purified by procedures outlined in Chapter 6 and the solution is accurately measured, evaporation of solutions on a strip of filter paper is recommended. This approach may be limited, however, to irradiations of short duration (5 min or less) because the paper may char during the activation. Filter paper of sufficient purity is extremely difficult to obtain.

4. Errors Due to Standard

In the procedures mentioned earlier for preparing standards, accurate results can be obtained if neutron self-absorption is absent from the sample

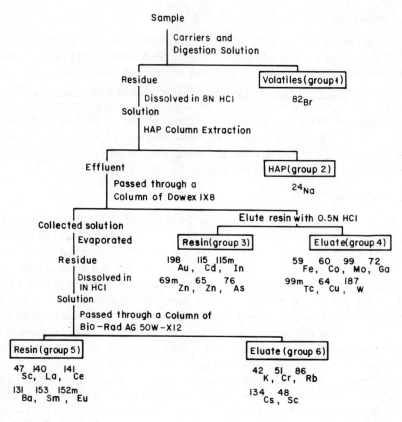

Fig. 1. Radiochemical group separations by ion exchange. Reprinted by permission from G. H. Morrison and N. M. Potter, *Anal. Chem.*, **44**, 839 (1972). Copyright by the American Chemical Society.

and the standard. Approximate calculations of the magnitude of this effect can be made.[16-18] When this effect cannot be avoided, the sample with a known addition of the desired element can serve as the standard. The same approach can be employed to determine the presence of self-absorption. For instance, we have doped solutions of sodium carbonate (0.5 g/ml) with 1.000 and 0.1000 μg/ml of Cu^{2+} and Mn^{2+}, respectively. Determinations of these elements in the solution and comparison with sodium-free solutions of the traces showed values within the expected experimental error of the doped levels. Samples of powdered sodium carbonate were subsequently analyzed with assurance that neutron self-absorption was negligible.

A compound or material that contains a known amount of the element to be determined is very suitable but rarely accessible. Fortunately advances in the production of standards for trace analysis have made available suitable matrices of animal and plant tissue, and of an inorganic material containing many trace elements (Chapter 2). For multielement analyses based on neutron activation, radiochemical group separations, and γ ray spectroscopy, such reference materials are helpful. The ion-exchange scheme outlined in Figure 1 has been used for processing SRM 1571, botanical leaves.[19] In this laboratory the solvent extraction scheme in Figure 2 was used to process SRM 610, a doped sodium-lime glass. Figure 3 gives γ-ray spectra of the various phases. For special analyses of water, air pollutants, biological fluids, foods, geological and lunar samples, and in forensic applications, such standards are valuable. However synthetic comparison standards will still be needed for many determinations by activation analyses.

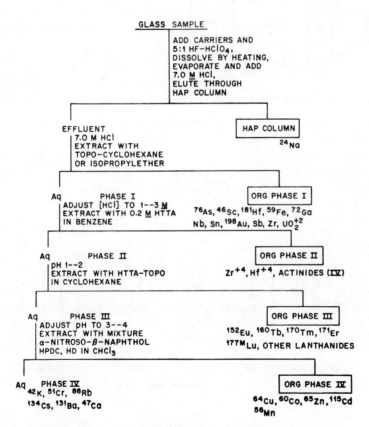

Fig. 2. Radiochemical group separations by solvent extraction.

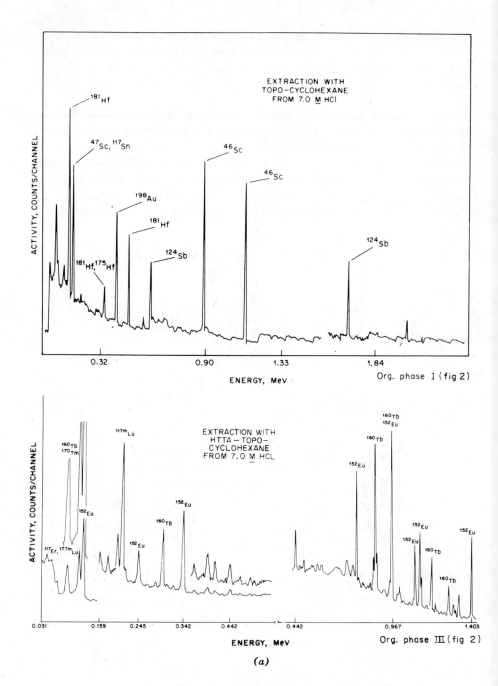

Fig. 3. γ-Ray spectra of fractions separated from glasses.

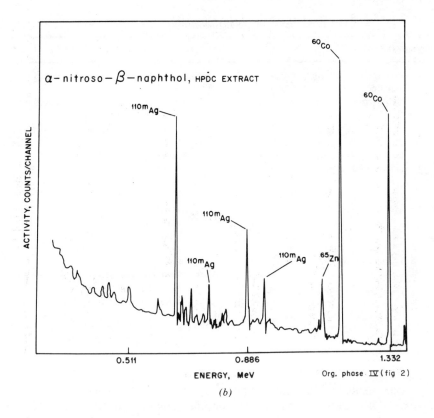

(b)

D. ACCURACY AND PRECISION

Table 5 outlines potential sources for systematic errors during activation analysis. The accuracy of the measurement depends largely on the reduction or elimination of these errors. Errors resulting from mechanical procedures described in items 1, 2, 3, 5, and 6a and 6b of Table 5 can be assessed individually. Sufficient knowledge of the elements likely to cause errors of the kind outlined in item 4 is needed to avoid these problems. Special procedures involving measurement of decay and absorption curves, comparison of relative ratios of activities from different photopeaks of an isotope, comparison of the energy and shapes of γ-ray spectra of preparations isolated from the sample and standard, and irradiations of samples with thermal neutron shields of cadmium may be needed to avoid errors of the kind mentioned in items 4a and 6b.

The precision of a series of measurements is principally a question of the random error involved. Precisions of ± 0.5–$\pm 20\%$ have been reported for

TABLE 5.   Systematic Errors During Activation Analysis

Source of Error	Procedures Involved
1. Contamination during preirradiation handling	Sampling, weighing, and packaging the sample
2. Comparison standard	a. Interfering radionuclide produced via reactions other than (n, γ)
	b. Radiochemical impurity, weighing or diluting errors
3. Loss of element by volatilization	Irradiation
4. Neutron flux attenuation or variation	a. Neutron self-absorption during irradiation
	b. Unequal flux at sample and standard
5. Losses	a. During recovery of sample after irradiation
	b. During chemical processing when carrier incorrectly added
6. Inaccurate counting	a. Counting geometry
	b. Corrections for decay
	c. Counting statistics
	d. Background and peak area determination
	e. Different attenuation of γ-ray in sample solution and in standard

measurements by activation. Where large total errors cannot be explained as a result of interference, the errors in the individual steps of the procedure should be determined. The error of the overall method can then be reduced by modifying the appropriate step. Large variations of repetitive measurements due to inhomogeneous distribution of the element in solid samples can be verified by dissolving the sample and analyzing aliquots of the solution. Fluctuations in other procedures—for example, the mean square deviations in weighing the sample and in the measurement of the activity of the sample—can be easily determined. After obtaining these data, the analyst assesses the error due to inhomogeneity of the neutron flux in the sample and standard, and errors in statistical fluctuation of the neutron flux. During nondestructive analysis of a series of liquids, one investigator found that error in weighing samples and standards was the most important factor affecting the precision of the determination.[7] In most cases where difficulties from competing nuclear reactions or neutron self-absorption are absent and enough activity is available to provide good counting statistics, precisions on the order of $\pm 4\%$ or better should be obtained routinely during the determination of ultratrace elements.

E. INTERFERENCES

Significant interferences in activation analysis are those resulting from the activation process. Because of the presence of a high, fast neutron flux at the core, (n, p), (n, α), and (n, 2n) reactions might well lead to the production of the same isotope of the element produced by (n, γ) reactions. When any two of these reactions generate the same isotope of the element in question, the phenomenon of interference occurs. Three main types of such interferences have been designated by Rakovic.[7] Primary interferences (Type A) occur when the interfering nuclide is formed by direct capture of the activating particle. Type B interferences are induced by a secondary

TABLE 6. Possible Interference Reactions in Thermal Neutron Activation

Isotope Measured	Nuclear Reactions Generating Same Isotope
^{24}Na	^{23}Na(n, γ)^{24}Na
	^{24}Mg(n, p)^{24}Na
	^{27}Al(n, α)^{24}Na
^{64}Cu	^{63}Cu(n, γ)^{64}Cu
	^{64}Zn(n, p)^{64}Cu
^{69m}Zn	^{68}Zn(n, γ)^{69m}Zn
	^{69}Ga(n, p)^{69m}Zn
	^{72}Ge(n, α)^{69m}Zn
^{75}Se	^{74}Se(n, γ)^{75}Se
	^{78}Kr(n, α)^{75}Se
^{28}Al	^{27}Al(n, γ)^{28}Al
	^{28}Si(n, p)^{28}Al
^{32}P	^{31}P(n, γ)^{32}P
	^{32}S(n, p)^{32}P
	^{35}Cl(n, α)^{32}P
	^{30}Si(n, γ)^{31}Si $\xrightarrow{\beta^-}$ ^{31}P(n, γ)^{32}P
^{14}C	^{13}C(n, γ)^{14}C
	^{17}O(n, α)^{14}C
	^{14}N(n, p)^{14}C
^{99m}Tc	^{98}Tc(n, γ)^{99m}Tc
	^{98}Mo(n, γ)^{99}Mo $\xrightarrow{\beta^-}$ ^{99m}Tc
^{47}Sc	^{46}Ca(n, γ)^{47}Ca $\xrightarrow{\beta^-}$ ^{47}Sc
^{56}Mn	^{55}Mn(n, γ)^{56}Mn
	^{55}Fe(n, p)^{56}Mn
	^{54}Cr(n, γ)^{55}Cr $\xrightarrow{\beta^-}$ ^{55}Mn(n, γ)^{56}Mn
^{76}As	^{75}As(n, γ)^{76}As
	^{74}Ge(n, γ)^{75}Ge $\xrightarrow{\beta^-}$ ^{75}As(n, γ)^{76}As

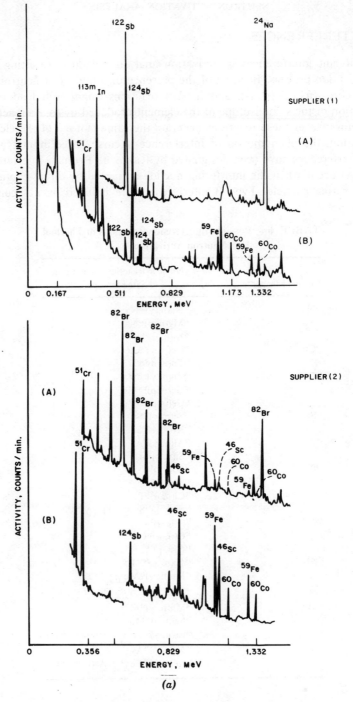

Fig. 4. γ-Ray spectra of silicon dioxides from various suppliers.

210

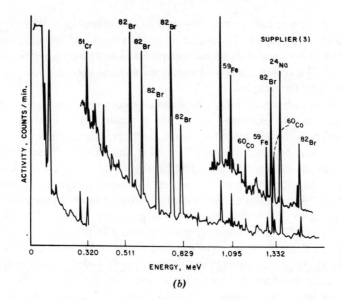

(b)

particle (a proton) that is ejected from a nucleus after the capture of the primary particle (neutron). The ejected secondary particle activates other nuclei to form an interfering radionuclide. The third type of interferences result from naturally radioactive nuclides of a given element. Type A interferences are the most significant with respect to thermal neutron reactions (n, γ). Some of the more common reactions generating the same isotope of an element are listed in Table 6. Standard procedures such as activating in the thermal column where the fast flux is negligible reduce Type A interferences. Careful examination of matrix components can detect the possibility of errors from Type B and C interferences.

F. ANALYSIS OF HIGH PURITY MATERIALS

Activation analyses have been applied successfully at many laboratories for the determination of traces in ultrapure materials. At Bell Laboratories silicon dioxides, specially fabricated for optical waveguide research, have been characterized by this technique.[20] The γ ray spectra of Figure 4 were obtained nondestructively to compare products supplied by several companies. Spectrum A of product 1 was recorded after a 24 hr cooling period. Considerable activity from ^{24}Na, ^{113m}In, and ^{122}Sb was detected. Spectrum B, obtained 2 weeks later by counting for 30 min, shows activity from ^{59}Fe and ^{60}Co at levels slightly above background. The spectrum of the material from supplier 2 was recorded 24 hr after irradiating the sample. The spectrum suggests that the sample was most likely produced by hydrolysis of SiCl$_4$, which was contaminated with SiBr$_4$. Photopeaks of ^{24}Na, ^{59}Fe, ^{51}Cr and ^{60}Co are also present. The spectrum of the compound from the third supplier contains isotopes of these same elements. The presence of ^{46}Sc [an (n, p) product of ^{46}Ti] in spectrum B of supplier 2 suggest that titanium impurities were present in the sample.

A series of samples of SiO$_2$ was simultaneously irradiated and compared with standards. An examination of data in Table 7 indicates that samples from supplier 5 contain less iron, cobalt, and chromium than materials supplied by others. No activity from ^{51}Cr, ^{59}Fe, and ^{60}Co was detected in samples of Suprasil (sample 6) after 100 hr of nuclear irradiation. Short-lived isotopes ^{64}Cu and ^{56}Mn were measured by γ-ray spectrometry following removal of ^{31}Si by volatilization with HF. Comparison of activation results with data from X-ray fluoescence and emission spectroscopy (Table 8) showed reasonable agreement, suggesting that the procedures used in the latter two methods were not significantly influenced by contamination.

Traces of copper and manganese in high purity sodium and calcium carbonates have also been determined.[21] In the case of sodium carbonate, the procedures involved a preliminary separation of ^{24}Na with a specific inorganic exchanger, hydrated antimony pentoxide, and a subsequent selective extraction of ^{56}Mn and ^{64}Cu with a mixture of pyrollidine carbodithioic acid and dithizone in CHCl$_3$. The same extraction process was used to separate these isotopes from the matrix isotopes ^{47}Ca and ^{47}Sc, produced by irradiating CaCO$_3$. Quantitative results for several determinations are reported in Table 9.

Glasses fabricated from these components have been irradiated with thermal neutrons and fingerprinted by γ-ray spectroscopy.[20] Important information has been obtained about the source and conditions under which several traces are introduced during the glass-melting operation. For

TABLE 7.   Trace Elements in Silicon Dioxide

| Supplier | Concentration^a (μg/g) | | | Other Impurities Detected |
	Fe	Co	Cr	
1	1.24	0.030	1.90	Na, Br
2	0.82	0.005	2.16	Na, Br, Sc, Hg
3	1.60	0.041	0.17	Na, Br, Sb
4	1.00	0.008	0.19	Na, Br
5	0.68	0.009	0.009	Na, In, Sb
6	N.D., <0.1	N.D., <0.003	N.D., <0.005	

^{a} Average of at least two determinations.

TABLE 8. Analysis of Silicon Dioxide

Element	Concentration (μg/g)		
	JTB[a]	BTL[b]	BTL[c]
Chromium	<0.01	0.008	0.0084
Cobalt	0.01	0.001	{ N.D., <0.001
Copper	0.01	0.03–0.09	{ 0.013 0.005
Iron	0.01	—	<0.100
Manganese	0.007	<0.01	{ 0.0085 0.0093
Nickel	<0.01	—	—
Vanadium	<0.01	—	—

[a] Emission spectrography, J. T. Baker Chemical Co.
[b] X-Ray fluorescence spectroscopy, Bell Laboratories.
[c] Neutron activation analysis, Bell Laboratories.

example, platinum impurities detected in a $CaCO_3$ sample were traced to the use of a platinum vessel during drying. Iridium and silver impurities detected in glass after 100 hr irradiations and 14 day decay periods were traced to a platinum vessel used in melting the raw materials for the glass. Comparisons of γ-ray spectra of glasses melted under oxidizing and reducing atmospheres have shown more container contamination from platinum and its impurities when oxidizing conditions are used.

Examinations of γ-ray spectra obtained under identical irradiation and

TABLE 9. Determination of Manganese and Copper in $CaCO_3$

Sample	Number of Determinations		Recovery (%)		Concentration Found (ng/g)	
	Mn	Cu	Mn	Cu	Mn	Cu
31 RD 327	4	5	83.8	95.3	31.5	<5
			47.0	82.8	58.3	<5
			44.9	72.7	44.5	<5
			71.0	72.7	40.0	<5
			—	86.3	—	1.6
32 RD 85	2	2	72	95	340 ±9	<5
			65	95		
32 RD 86	2	2	72	95	658 ±13	<5
			79	95		

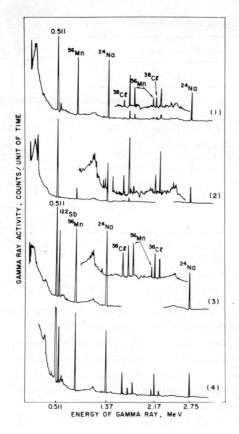

Fig. 5. γ-Ray spectra of nitric acids from commercial suppliers. Reprinted by permission from J. W. Mitchell, C. L. Luke, and W. R. Northover, *Anal. Chem.* **45,** 1503 (1973). Copyright by the American Chemical Society.

counting conditions permit a facile screening of several high purity reagents sold by commercial distributors. Visual inspection of the photopeak heights of ^{24}Na, ^{56}Mn, and ^{38}Cl (Figure 5) permits a convenient comparison of the relative purity of nitric acids with respect to these traces. Quantitative results for the analysis by neutron activation of a number of commercially available high purity reagents were reported previously.[22]

The practical use of neutron activation for trace determinations and characterizations of high purity materials is increasing annually. Its increasing importance is implied by the topic discussed at a recent symposium: "Is Radioactivation the Ultimate in Trace Analysis?"[23] Comprehensive reports on activation analysis are numerous.[24-30]

II. RADIOISOTOPE DILUTION (RID)

A. INTRODUCTION

Recently, radioisotopes have shown practical utility for the determination of trace elements.[31, 32] The pioneering work of Ružička and co-workers[33-37] clearly illustrates the advantage and utility of this approach. Their technique is based on the substoichiometric principle and in some cases it approaches the sensitivity of neutron activation, as Table 10 indicates. The method is potentially broadly applicable as revealed by the list of elements with isotopes suitable for determination by radioisotope dilution (Table 11). Because of the high precision and accuracy and low cost of the method, applications in trace analysis could increase considerably as expertise in ultrapurity techniques becomes widespread.

B. QUANTITATIVE PRINCIPLES

Quantitative elemental analysis by isotope dilution depends on the reduction in the specific activity of a radioisotope when it is mixed with nonactive atoms of the element in the same chemical state. The specific activity of the tracer solution S^*, defined as the ratio of the total activity A^* to the total quantity of element W^* in the tracer solution, is represented by

$$S^* = \frac{A^*}{W^*} \tag{4}$$

When the tracer is added to a solution containing the amount W of the ele-

TABLE 10. Reported Substoichiometric Isotope Dilution Methods

Isotope	Amount Determined by Isotope Dilution[a]	Interference-Free Sensitivity of Neutron Activation	Relative Error Isotope Dilution (%)
^{65}Zn	1.00×10^{-9} g/ml	1.0×10^{-8} g	1.0
^{64}Cu	2.4×10^{-8} g	2.0×10^{-10} g	1.7
^{203}Hg	10^{-8} to 10^{-9} g/ml	8×10^{-9} g	2.5
	10^{-8} to 10^{-9} g/ml		28
^{59}Fe	3.0×10^{-8} g	2×10^{-7} g	8.5
^{114m}In	5.5×10^{-11} g	6.0×10^{-11} g	45
^{171}Tm	5.0×10^{-7} g	2×10^{-7} g	—
^{195}Au	1.0×10^{-8} g	5×10^{-11} g	30
^{131}I	9×10^{-7} g/ml	2×10^{-8} g	17
^{88}Y	5.8×10^{-7} g	4×10^{-7} g	—

[a] Regarding items 1 and 4, this column, see Devoe.[31,32] Other data from Ružička and co-workers.[33-38]

TABLE 11. Elements with Suitable Radioisotopes for Determination by Isotope Dilution

Antimony	Indium	Selenium
Beryllium	Iodine	Silver
Bismuth	Iron	Sodium
Cadmium	Lead	Strontium
Calcium	Manganese	Sulfur
Cerium	Mercury	Thulium
Cesium	Niobium	Tin
Chromium	Phosphorus	Titanium
Cobalt	Rare earths	Tungsten
Copper	Ruthenium	Vanadium
Gallium	Samarium	Yttrium
Gold	Scandium	Zinc

ment in the nonradioactive form, equation 4 becomes

$$S = \frac{A^*}{W^* + W} \tag{5}$$

where S is the specific activity of the mixture. By combining equations 4 and 5, the expression

$$W = W^* \left(\frac{S^*}{S} - 1 \right) \tag{6}$$

is obtained. The unknown amount W of the nonradioactive element in a sample can be quantitatively measured by isolating a known quantity of the element from the mixture of sample and standard radiotracer, and measuring the resulting activity. Since quantitative results depend only on a measurement of the specific activity, it is not necessary to recover the entire quantity of the substance from the sample solution. This constitutes one of the greatest advantages of analysis by isotope dilution—namely, the ability to provide quantitative results without quantitatively recovering the entire amount of the material initially contained in the sample. However to ascertain the specific activity of the recovered fraction, the quantity of sample isolated must be determined.

During the first few years after the method was developed, the specific activity was found primarily by measuring the activity of a known weight of the element isolated as a precipitate; however any suitable analytical method can be used. Because the amount of the isolated element must be known accurately, early isotope dilution methods were restricted to the determination of quantities of constituents in samples easily measured by weighing the precipitate ($\geq$ 10 mg). As other more sensitive analytical methods were developed, the use of isotope dilution procedures for inorganic analyses declined rapidly.

C. SUBSTOICHIOMETRIC RADIOISOTOPE DILUTION (SRID)

Ružička and Starý devised a new procedure for isotope dilution analysis that extended the application of the method into the submicrogram region.[10] Their method eliminated the need to determine the specific activity of the recovered fraction.

Upon substitution into equation 6 for the specific activity of the standard radioisotope and sample solutions, one obtains

$$W = W^* \ [(a_1^*/w_1^*/a_2/w_2) - 1] \tag{7}$$

where a_1^* and a_2 are the measured relative activities emitted by the quantities w_1^* and w_2 of the element isolated from the solutions containing the standard (radiotracer) and from the mixture of the radiotracer with the sample, respectively. If the same weight of material is isolated from the standard and sample solutions, equation 7 reduces to

$$W = W^* \left(\frac{a_1^*}{a_2} - 1\right) \tag{8}$$

which shows that an element in a sample can be determined by measuring the relative activities of equal quantities of material isolated from the standard and sample solutions. The factors that limit the sensitivity now become the detection limit for the measurement of the activity and the smallest quantity of the material that can be reproducibly isolated from the solutions.

Solvent extraction provides an effective means of isolating small amounts of various elements. The general reaction for the solvent extraction of a cation is given by

$$M_{aq}^{n+} + nHA_{org} \rightleftharpoons MA_{n \ org} + nH_{aq}^+ \tag{9}$$

In a typical procedure a large excess of the chelating agent HA completely complexes M^{+n}, the cation of charge n, and produces a quantitative transfer of the metal into the organic phase in the form of the chelate complex MA_n. In contrast to this approach, a substoichiometric amount of the chelating agent can be used to transfer only a portion of the cation into the organic phase. Under appropriately chosen conditions that ensure 99% or greater reaction of the chelating agent, a solvent extraction process can be used to isolate reproducibly equal quantities of a cation from solutions that initially contain different amounts of the element. Substoichiometric amounts of powerful chelating agents are necessary for these separations.

A substoichiometric amount of a multidentate chelating agent such as EDTA can also be used to isolate a quantity of cation by forming an anionic chelate complex. Upon passage of the solution through a cation exchange resin in the sodium form, the excess cation is retained on the column and the metal complex is eluted. Appropriate conditions for substoichiometric isolations can be chosen by examining stability and equilibrium constants.

An added advantage of the substoichiometric approach is that solvent extraction becomes more selective in the separation of cations than it is in the normal extraction process in which a large excess of the chelating agent is employed.[38]

D. SRID VERSUS NAA

Although neutron activation is the more widely used nuclear method, radioisotope dilution compares favorably in several categories.

1. The cost of radioisotopes (available commercially as carrier-free or in high specific activity form) and of counting equipment is within the budget of most industrial and academic laboratories. Neutron generators, reactors, and service irradiations are costly.

2. Low levels of radioactivity must be handled carefully in radioisotope procedures, but careful execution of procedures is an essential part of analytical work. Highly radioactive samples that may result from activation can require elaborate shielding and mechanical manipulators to avoid radiation hazards.

3. Substoichiometric isotope dilution, although limited to the accurate determination of a single element, is potentially useful for measuring 10^{-5}–10^{-11} g of each of 36 elements listed in Table 11. Neutron activation can be used to determine more than 65 elements, either individually or collectively by radiochemical group separations at 10^{-6}–10^{-12} g. The latter method is also convenient for semiquantitative survey analyses.

In cases of elements with very short half-lives or those with insufficient sensitivity for thermal neutron capture, the availability of suitable radioisotopes makes SRID superior to NAA. The determination by isotope dilution of iron with ^{59}Fe and beryllium with carrier-free ^{7}Be are excellent examples.[39] Similarly, lead cannot be determined by thermal neutron activation analysis, but ^{210}Pb (RaD) prepared from radium ore, can be used for substoichiometric determinations of this element by isotope dilution.

Isotope dilution analyses are more difficult to perform than measurements by activation, since it is necessary to minimize or eliminate all possible sources of contamination throughout the procedure. Reducing contamination and eliminating losses of the element from solution become the main problems in SRID. To keep the blank at a level that should be at least 10 times lower than the determined amount of the test element, the general methods described previously for eliminating contamination during sampling and dissolution must be adopted.

The selective separations on which SRID methods are based make it possible to calibrate and determine traces in the most diversified matrices. However perfecting these procedures takes several months to a year. Fortunately losses during the separation procedure can be easily detected by counting beakers, flasks, columns, and other vessels that are used.

E. GENERAL PROCEDURE

The following is a general procedure for trace analysis by isotope dilution.

1. Add a known quantity of the standard radioisotope to the sample and dissolve the mixture under conditions that prevent contamination. The specific activity of the radiotracer is determined most conveniently by reverse isotope dilution.

2. In the same manner used for the sample solution, chemically treat a blank solution containing the same amount of radioisotope used in step 1.

3. Substoichiometrically isolate equal amounts of the element from both solutions and measure the relative activities of the isolated fractions under identical counting conditions. Specific procedures for determining various traces are given by Ružička et al.[38]

Although the method is simple in principle, to obtain reliable data the following details must receive appropriate attention:

1. All reagents and water must be specially purified by methods discussed in Chapter 5.

2. Contamination must be eliminated from airborne particulate matter by working in all-plastic, positive-pressure clean hoods.

3. Standard radiotracer solutions are to be prepared immediately before use by spiking nonactive solutions of known concentration with the carrier-free isotope and diluting to the desired concentration. Direct counting or reversed isotope dilution analysis of the diluted solution gives a precise determination of the quantity of element in an aliquot of the standard.

4. Appropriately cleaned, leached, and equilibrated polyolefin or Teflon vessels must be used.

5. Adsorption phenomena become particularly important because the

total amount of element being measured is in the nanogram range. Losses to the container surface must be evaluated for a given element. As a general rule the solution should be transferred the least number of times in the smallest possible containers (i.e., minimize surface to sample volume ratio).

6.   Complex formation with micromolar solutions occurs more slowly than when macromolar concentrations are used. Reaction times must therefore be determined and repeated precisely during subsequent analyses.

F.   APPLICATIONS

Any element with a suitable isotope or organic compound with a labeled atom can be measured by radioisotope dilution. Such diversified analyses as the macro determination of the γ-isomer of benzene hexachloride by ^{14}C isotope dilution,[40] of halogen-containing carboxylic acids and anhydrides by [36]Cl,[41] and of amino acids,[42] as well as submicrogram determinations of element traces,[36, 37] have been accomplished.

Measurements of the carrier content of radioisotopes by reversed substoichiometric isotope dilution are particularly useful and must be performed frequently, since the specific activity of commercially available products is often specified as a range, for example (5–10 Ci/g). Procedures for the determination of many traces in various ultrapure alloys and materials have been described.[38]

III.   STABLE ISOTOPE DILUTION (SID)

A.   GENERAL

Mass spectrometric analysis by isotope dilution is a powerful tool well suited for the determination of all but a few elements (ca. 20) that occur monoisotopically in nature. All the other multi-isotopic elements are now available at reasonable cost from the Isotope Development Center, Oak Ridge National Laboratory. The major advantages of the method are identical to those just enumerated for radioisotope dilution—high accuracy (better than 1% for most elements and with some at the 0.1% level), and the possibility of obtaining quantitative results without complete separations or recoveries. Because chemical processing of the sample and selective separations are required, the blank problem, common to all of the trace methods except activation analysis, is the primary factor limiting the sensitivity of the method.

B.   BASIS FOR QUANTITATIVE ANALYSIS

If one of the two lightest isotopes of the element undergoing measurement is available in the enriched form, it can usually be intimately mixed with a sample containing the element in its naturally occurring state without any chemical differences between the isotopes. When the sample is introduced into a mass spectrometer, the total detector signal from the lightest isotope $S_{m_1}^t$ is given by

$$S_{m_1}^t = S_{m_1}^n + S_{m_1}^e \tag{10}$$

where the signal from the lightest isotope in the sample is proportional to the number of atoms of M_1^n occurring naturally. The relationship

$$M_1^n = A^n M_t \tag{11}$$

shows that the number of atoms of the lightest isotope in the sample is given by the product of its isotopic abundance A_1^n and the total number of atoms M_t of the element in the sample.

The intensity ratio R of the two isotopes M_1 and M_2 of the element can be written as

$$R = \frac{A_2^n M_t^n + A_2^e M_t^e}{A_1^n M_t^n + A_1^e M_t^e} \tag{12}$$

where A_1^e and A_2^e are the isotopic abundances of the two isotopes in the enriched stable isotope solution and M_t^e is the total number of atoms of the two isotopes in the enriched solution. Equation 12 can be rearranged to

$$M_t^n = M_t^e \left\{ \frac{A_2^e - R A_1^e}{R A_1^n - A_2^n} \right\} \tag{13}$$

By expressing equation 13 in terms of weights instead of atoms, the expression

$$W_t^n = W_t^e \left\{ \frac{\text{aw of } M^n}{\text{aw of } M^e} \right\} \left\{ \frac{A_2^e - R A_1^e}{R A_1^n - A_2^n} \right\} \tag{14}$$

can be written, where aw is the atomic weight. The unknown weight of the element in the sample W_t^n is determined by adding a known weight of the enriched isotope W_t^e and subjecting the sample to whatever chemical steps are necessary to give a homogenous mixture of the isotopes. The ratio R of the abundance of the two isotopes is measured, and equation 14 is used to determine the weight of element initially present in the sample.

C. EXPERIMENTAL PROCEDURE

The procedure for stable isotope dilution involves essentially five steps.

1. A weighed portion of the standardized enriched isotope solution is added to a known weight of the sample. This spiking procedure has been accomplished by filling a polyethylene or Teflon syringe with the desired

amount of isotope solution, removing the platinum or stainless steel needle, and capping the tip with a Kel-F cap. After dissipating any static charge by wiping the exterior of the syringe with a damp cellulose towel,* the syringe and contents are weighed on a semimicro balance to 0.02 mg. To a beaker containing the material to be analyzed is delivered a portion of the spike solution from the syringe. After reweighing the syringe, the weight of the spike solution can be determined by difference with less than 0.01% uncertainty.[43] Multiple spiking of a single sample for the determination of many elements can be performed.

2.   The sample is dissolved by appropriate methods that ensure thorough mixing of the solution to achieve equilibrium of the enriched isotope with the element to be measured. Several blanks are prepared by adding the same amount of the spike solution to beakers without samples and treating these in the same manner as the sample. Since submicrogram quantities are being measured in the sample, great care must be exercised to avoid contamination. The procedure for preparing the sample must therefore involve a minimum number of handling steps. Only a few chemicals in very small amounts should be used, and the wall surface of containers must be reduced to a minimum. The high sensitivity of the method easily allows the level of contamination introduced by the entire procedure to be determined by analyzing spiked blanks.

3.   The isotopically altered element must be chemically separated from the sample. Any suitable separation technique is satisfactory. For example, 1 ng each of thallium and lead have been electrolytically deposited on platinum anodes as Tl_2O_3 and PbO_2 with better than 70% recovery.[44] At the same time copper and silver were deposited on the cathode. Rubidium and potassium were measured after separation by precipitation with the acid form of tetraphenyl boron and recovery of the free cations by ion exchange.[43]

Substoichiometric separations based on electrochemical deposition have greatly simplified procedures for stable isotope dilution and are preferable to other methods.[45]

4.   The isotopic ratio of the element isolated in the separations step is measured by mass spectrometric methods.

5.   The amount of impurity element in the sample is then calculated from equation 14.

It is necessary to determine the ratios of A_2/A_1 for both the enriched isotope and for the pure natural isotope to detect any discrimination as a result of detector nonlinearity, variations in sensitivity as a function of

---

* See Chapter 6 for instructions on purifying filter paper (or paper towels).

mass, or systematic errors in the measurement of peak areas. The relative isotopic abundances of the element being determined must also be measured to ensure that no deviations from the expected natural abundance has resulted during the history of the sample.

1. Calibration of Isotope Solutions

Investigators seldom have available sufficiently accurate information on the abundance or purity of enriched isotope solutions. The concentration of the enriched isotope thus must be determined by the addition of a known amount of the natural element to a weighed portion of the enriched (spike) solution. The isotopic abundance ratio of the element is determined by mass spectrometry, and the concentration of the enriched solution is calculated by use of equation 14.

Proper storage of the calibrated solution is essential. Storage in borosilicate glass flasks specially modified to prevent evaporation or in Teflon bottles is recommended.[43] Glass flasks have been stored in desiccators at 100% relative humidity with no detectable change in weight of the solution over a 1 year period. Storage in Teflon bottles was found to be less satisfactory (loss < 0.1% over 3 months), but this is necessary when the enriched isotope solution could be contaminated by borosilicate glass.[43]

2. Recent Innovations

Mass spectrometers with surface ionization by thermal means have been used widely for solid samples, and instruments with electron impact sources have been employed for the determination by isotope dilution of constituents in gases. Variable sensitivities for different elements, one limitation of the thermal source instrument, has been largely overcome by using spark source mass spectrometers.[46]

A recent innovative approach, metal-labeled stable isotope dilution (MLSID), solves the problems of volatility, separation, and detection by converting the element to be determined to a volatile metal chelate. This eliminates the need for thermal or spark sources for nonvolatile elements. Now medium-resolution mass spectrometers ($M/\Delta M$ = 5000), which are available in many analytical departments, can be used to determine many metals using the ordinary electron impact instrument. Another important advantage of this method is the elimination of matrix effects. However the preparation of the volatile compound is difficult to accomplish in some cases.

Since the mass spectral detection limits for several metal chelates are very high[47] as Table 12 reveals, extremely sensitive stable isotope dilution

TABLE 12. Mass Spectrometric Sensitivities for Metal Chelates

Element	Chelating Agent[a]	Quantitative Results (g)	Sensitivity	References
Aluminum	H(fod)	10^{-12}	10^{-14}	47
Thalium	H(fod)	10^{-12}	10^{-15}	47
Iron	H(fod)	10^{-12}	10^{-15}	47
Chromium	H(fod)	—	10^{-14}	47
Nickel	H(fod)	—	—	47
Copper	H(fod)	—	—	47
Yttrium	H(fod)	—	—	47
Palladium	H(fod)	—	—	47
Rare earths	H(fod)	—	—	47
Lead	H(fod)	—	—	47
Holmium	H(fod)	10^{-6}	10^{-10}	47–49

[a] 1,1,1,2,2,3,3-Heptafluoro-7,7-dimethyl-4,6-octanedione.

methods can be developed. A successful method for the determination of zirconium and chromium in geological samples has been reported.[48, 49]

3. Interferences

Interferences in stable isotope measurements result primarily from isobars present in the sample and from multiply charged species. Knowledge of traces and major constituents in the sample aids in determining possible interferences. When applicable, spark source mass spectroscopy is an excellent tool for providing such broad spectrum survey analyses of the sample.

D. ANALYSIS OF ULTRAPURE REAGENTS

Highly pure acids prepared by subboiling distillation have been analyzed at NBS by isotope dilution.[50] In this work 100 ml samples of water and various acids were spiked with 1 μg/ml isotopic solutions freshly prepared by diluting 50 μg/ml master spike solutions. Highly pure acids were spiked for concentrations of 1–0.1 ng/g. Reagent grade chemicals were spiked in the 1–100 ng/g range. Suitably spiked samples were then concentrated in the evaporator shown in Figure 6. This apparatus was devised to provide a clean environment for evaporating a spiked sample and for transferring the residual volume onto high purity gold wires. A clean atmosphere was provided by filtering nitrogen through a 0.1 μm cellulose filter. The flowing nitrogen swept out the evaporating acids and prevented the entry of particulate contamination.

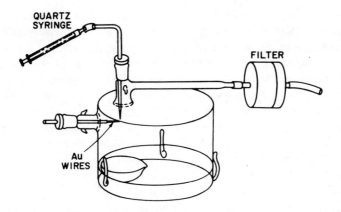

Fig. 6. Quartz evaporation chamber (NBS). Reprinted by permission from E. C. Kuehner et al., *Anal. Chem.*, **44**, 2050 (1972). Copyright by the American Chemical Society.

A pure vitreous silica pipet was used to transfer the last few drops of acid onto the gold wires (shown positioned from the side joint). Partial loss or incomplete transfer of the evaporated sample to the gold wires was ignored, since quantitative analytical data were only a function of the altered isotopic ratio. It sufficed to transfer onto the wires enough element for adequate measurement.

The gold wires were sparked in a spark source mass spectrometer to obtain a graded series of exposures on photographic plates. Seventeen elements were simultaneously determined at the fractional nanogram per gram level, with uncertainties ranging from ±10 to 30% depending on the value of the altered ratio.

Several high accuracy methods for the determination of traces by isotope dilution have been reported. Thorium, uranium, lead, and thallium in silicate glass standards were measured by SID at 20 ng/g with precisions of the order of 0.5%[51] A method for the anodic deposition and determination of 10 μg to 10 ng of lead by isotope ratio mass spectrometry has been described.[52] Traces of silver, copper, molybdenum, and nickel in ingot iron were measured by spark source mass spectrographic isotope dilution.[53]

IV. X–RAY FLUORESCENCE SPECTROSCOPY

A. INTRODUCTION

Attempts by several workers to obtain quantitative results during the early development of X-ray analytical methods met with difficulties because the approach involved directly exposing the untreated sample to a suitable excitation source.[54-57] Although reasonably high sensitivity could be obtained in this way, the intensity of the fluorescent X-radiation was a complicated function of the matrix. Additionally, the intensity or the energy of the emitted X-ray was influenced by absorption and enhancement effects from elemental interactions, physical effects resulting from variations in particle size and surface, and sometimes the chemical state of the element. Since the intensity of the characteristic X-radiation of a given wavelength is a function of the number of atoms of the corresponding element present in the portion of the sample that is emitting radiation, the observed intensity of an analytical line of an element in a sufficiently thin film can be related to the weight of that element present in the sample.

Efficiencies of the various excitation, absorption, scattering, diffraction, and detection phenomena are not known accurately enough to permit direct computation of concentrations or amounts from intensity data. It is necessary to compare the intensities of unknowns with those from known standards, to obtain quantitative information. Thus standards must be prepared with great care to ensure either the same particle size distribution for powders or the same surface characteristics for alloys or solids. For reliable results, standards must be as similar as possible to the unknowns in composition, as well.

Matrix effects were often such that wide deviations from linearity resulted when standards covering a concentration range of an order of magnitude or greater were measured. Satisfactory resolution of difficulties due to particle size and density variations were accomplished by high pressure pelleting of the sample.[58] Absorption and ehancement effects have also been compensated for by internal standard techniques.[59] Attempts to apply mathematical matrix correction procedures[60-62] and empirical parameters relating concentration and line intensity[63] have met with some success but do not solve the problem of particle size contributions.

Advances in the development of high intensity sealed X-ray tubes and improvements in the sensitivity of radiation detectors have been responsible for a rapid growth of X-ray spectrometry in recent years. The method has become a major instrumental technique because of speed, convenience, simplicity, and easy adaptability for semiautomatic nondispersive analysis. Qualitative and quantitative determination of all elements heavier than atomic number 11 (sodium) is possible for major constituents (0.1–100%) in a sample.

B. SEPARATIONS IN X–RAY ANALYSIS

Several investigators have attempted to circumvent matrix effects by isolating trace elements from the matrix components before X-ray measurement.[64, 65] Luke developed such an approach for the determination of 69 of the 72 elements customarily determined with an X-ray spectrograph.[66] By combinining separation and preconcentration steps, this investigator extended the application of the X-ray method to the determination of trace elements (0.1–50 μg). Sensitivities for elements by this method are given in Table 13.

In Luke's Coprex method, the sample is dissolved in dilute acid solution and freed from interfering elements by making preliminary chemical separations by adding a suitable element (50–200 μg) to act as a coprecipitant. The trace elements are precipitated under proper conditions with a suitable organic or inorganic precipitant. The precipitate is filtered on a small paper disk, washed and dried, and the X-radiation of the trace

TABLE 13. Sensitivities for Various Elements by Coprex[a]

Sensitivities				
1–5 μg		5–20 μg		20–50 μg
Ag	Li[b]	F[b]	Tm	Al
B[b]	Mn	Ga	W	Au
Ba	Ni	Cd	Y	Bi
Be	P	Ge	Yb	Ir
Ca	S	Hf	Zr	Mo
Co	Sb	Hg		Na[b]
Cr	Sc	Ho		Pb
Cs	Sn	I		Pt
Cu	Tb	La		Si
Fe	Ti	Lu		Ta
In	V	Nb		U
K	Zn	Nd		
	As	Pr		
	Br	Rb		
	Cd	Rh		
	Ce	Se		
	Cl	Sm		
	Dy	Sr		
	Er	Te		
	Eu	Th		
		Tl		

[a] Data reported by S. M. Vincent and J. E. Kessler, Bell Laboratories Technical Memorandium, 1970. Minimum amounts required for quantitative determination with a precision of ±2% of the amount present.
[b] Indirect determination.

elements in the precipitate are measured on a conventional flat crystal spectrograph.

The main advantages of this method (Coprex) are (1) the elimination of line interferences from matrix elements or from other trace elements, (2) the determination of microgram level trace elements by using a suitable-coprecipitating element and reagent, and (3) the removal of interelement absorption and enhancement effects by isolating the trace elements uniformly dispersed in a low atomic number (organic) matrix.

The most useful broad spectrum reagents for Coprex have proved to be sodium diethyldithiocarbamate and ammonia. These two reagents, in the presence of suitable coprecipitants, are used to recover traces of the elements charted in Figure 7.

Precipitation from a solution buffered at pH 4.0 with acetate was effected by adding 50 μg of copper as the coprecipitant and a freshly prepared aqueous solution of 2% carbamate. At pH 9, the elements are precipitated

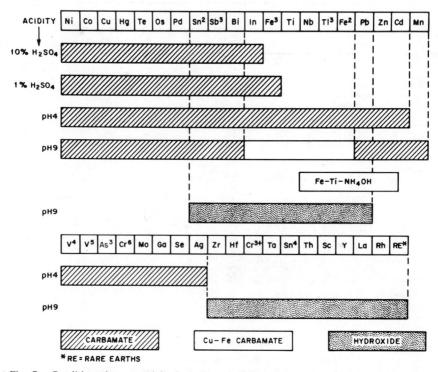

Fig. 7. Conditions for coprecipitation of trace elements with carbamate and ammonium hydroxide. Reprinted by permission from C. L. Luke, *Anal. Chim. Acta*, **41**, 242 (1968). Copyright by Elsevier Scientific Publishing Company.

with carbamate in the presence of 50 μg of Fe (III) and Cu (II). When precipitation of only the hydroxide elements is desired, 100 μg of Fe (III) is used as the coprecipitant and the solution is filtered immediately after neutralization to pH 9.

Selective separations of groups of elements are possible by controlling the pH at which precipitation occurs. Figure 7 shows that the selectivity of carbamate increases with acidity. Thus the first 11 elements can be separated from hydroxide elements, the alkalies, the alkaline earths, and from most of the platinum metals. In this way traces can be separated from alloys. For example, it has been demonstrated that traces of the 11 elements can be recovered from as much as 1 g of manganese in 1% sulfuric acid. Conditions for the selective precipitation of various elements have been described by C. L. Luke.[55]

C. MICROANALYSIS

To adapt the flat crystal spectrograph for microanalysis, either the excitation beam from the x-ray tube or the fluorescent beam from the sample must be apertured. Although the desired area of the sample is isolated and background radiations from surrounding material are eliminated, a reduction in intensity of the desired radiation parallels a decrease in the size of the sample or a reduction in the dimensions of the aperture. Since the intensity obtained with a flat crystal spectrometer is directly proportional to the cross-sectional area of the beam, only a limited reduction in sample size can be made before X-ray intensities become too low for practical analysis.[67]

X-Ray spectrometers for microanalysis have been perfected by improving the efficiency of secondary X-ray utilization by designing optical systems capable of accepting divergent rather than only parallel X-rays. Small samples or selected parts of a material are then analyzed by using an aperture to reduce the X-ray beam, and a sample holder constructed like a microscope stage. This instrument, capable of probing areas whose dimensions are measured in millimeters, is called the "X-ray milliprobe."[67]

The milliprobe constructed at Bell Laboratories is an adaptation of the classical Johannson arrangement. (Fig. 8) The advantage of this spectrometer results from a system of precision bar links rather than special highly precise gears or cams. It also covers a wider angular range than most of the mechanisms used previously and provides a convenient means for choosing the compromise between intensity and wavelength resolution most suitable for a particular analytical problem.[67]

Coprex techniques for the determination of nanogram quantities of trace elements have been developed by suitably confining the precipitate to a 2.54

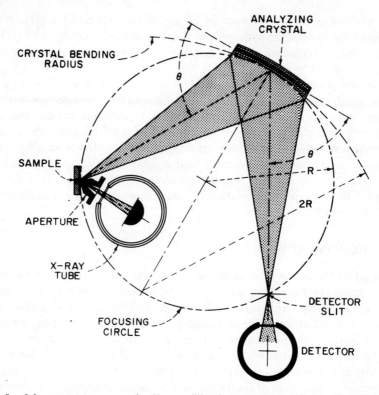

Fig. 8. Johannson arrangement for X-ray milliprobe spectrometer. *Source*: Ref. 67. Copyright 1967, Bell Telephone Laboratories, Incorporated. Reprinted by permission, Editor, Bell Laboratories RECORD.

mm spot and subsequently analyzing the precipitate in a curved crystal X-ray milliprobe spectrometer. The original apparatus for collecting the precipitate has been described by Luke.[66] A recently modified unit for collecting precipitates or insoluble particulates in a circular area with a 2.54 mm diameter was devised by J. E. Kessler and S. M. Vincent[68] (Figure 9). Further improvements in design have resulted in devices for pressure filtration of the solution in a syringelike unit that confines the precipitate to a 1.27 mm spot.

D. ANALYSIS OF PARTICULATES

Airborne particulates containing high concentrations of cations represent a potential source of significant contamination of pure solid reagents.[69] Although the concentration of particulates in inorganic salt solutions and the

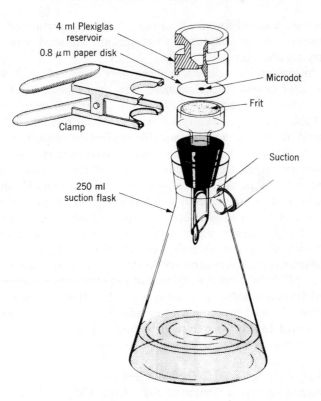

Fig. 9. Filtration apparatus for Coprex method.

efficiency of their removal by filtration have been determined by light scattering techniques, this method does not allow the elemental composition of the particles to be determined.[70] An extremely sensitive X-ray fluorescence method has been devised in one of the authors' laboratories for the direct determination of cations in particulates separated from commercially available ultrapure salts.[71]

After 0.5 g of an ultrapure salt is dissolved completely in 2.0 ml of deionized quartz-distilled water in a 10 ml Teflon (TFE) beaker, the solution is filtered through a 0.8 μm Millipore disk (cellulose acetate, 25 mm in diameter) in the apparatus of Figure 9. To prevent airborne contamination of the sample during these procedures, the reagent was opened, sampled, and weighed in the clean hood. After the dissolved sample was filtered and the insoluble particulates were washed with a few drops of pure water, a gray or black spot was found on the filter disk. The disks were dried and the microdots were counted for 40 sec, using a 1.25 mm aperture on a curved-crystal, vacuum, X-ray milliprobe spectrograph, equipped with a lithium fluoride crystal and a tungsten target.

X-Ray intensities from the paper disks were measured at characteristic wavelengths for copper, nickel, cobalt, iron, manganese, chromium, and vanadium. After blank corrections from the paper disk were applied (see blank 2, Table 14), the concentration of each metal present in the sample was obtained by comparison with suitable calibration disks. Previous work demonstrated that no significant uncertainties arise from the comparison of X-ray intensities from trace metals in particulates on filters with corresponding trace metals precipitated and collected on calibration disks.[72]

Calibration disks were prepared as follows: 1 ml of pure water, a drop of 0.01% aqueous solution of meta-cresol purple, 0.2 ml of 10% HCl containing 10 μg of Ti(IV) per milliliter, and an aliquot of a solution containing 100 ng of each of the seven metals were added to a 10 ml Teflon beaker. The resulting mixture was neutralized to the distinct purple color of the indicator (i.e., pH 9) by dropwise addition of pure isopiestically prepared ammonium hydroxide solution. Five drops (ca. 0.25 ml) of a 2% solution of sodium diethyldithiocarbamate was then added. The mixture was swirled, covered, allowed to stand 5 min, and filtered. The precipitate was washed and then treated as described earlier. To provide for background and reagent blank correction, a second disk was prepared in which the addition of the 100 ng of the seven metals was omitted (see Table 14).

The precipitation and collection of Fe^{3+}, Co^{2+}, Cu^{2+}, Cr^{3+}, and Mn^{2+} on calibration disks were monitored by radioisotope techniques. Triplicate measurements showed 98.5 $\pm$0.9, 83.6 $\pm$2.4, 94.0 $\pm$1.1, 98.6 $\pm$0.9, and 99.5

TABLE 14. Calibration Data for X-Ray Analysis

Cps	Metals						
	Cu	Ni	Co	Fe	Mn	Cr	V
Total cps[a]	1131 $\pm$20	1087 $\pm$45	634 $\pm$50	901 $\pm$8	393 $\pm$25	441 $\pm$31	480 $\pm$31
Blank 1[b]	631 $\pm$47	233 $\pm$2	84 $\pm$3	340 $\pm$14	63 $\pm$3.	76 $\pm$0	372 $\pm$10
Net cps	500 $\pm$43	854 $\pm$43	550 $\pm$47	555 $\pm$20	331 $\pm$28	365 $\pm$30	108 $\pm$37
Blank 2[c]	225 $\pm$14	63 $\pm$4	32 $\pm$1	83 $\pm$28	27 $\pm$2	36 $\pm$2	15 $\pm$1

[a] Average of values from two calibration disks obtained during a counting period of 40 sec.
[b] Average value obtained by counting two blank calibration disks.
[c] Average of values from four blank sample disks.

Reprinted by permission, J. W. Mitchell, C. L. Luke, and W. R. Northover, *Anal. Chem.*, **45**, 1506 (1973). Copyright by the American Chemical Society.

TABLE 15. Determination of Impurities in Particulates Separated from Ultrapure Salts

Sample	Concentration of Trace Elements[a] (ng/g)						
	Cu	Ni	Co	Fe	Mn	Cr	V
Na_2CO_3[b]	48	134	152	436	89	23	83
Na_2CO_3	N.D.	4	3	70	N.D.	N.D.	N.D.
NaCl	13	1	N.D.	16	N.D.	N.D.	N.D.
KI	N.D.	6	31	54	N.D.	N.D.	N.D.
NH_4Cl	N.D.	2	2	23	4	5	N.D.
NaAc	N.D.	N.D.	N.D.	64	N.D.	N.D.	N.D.

[a] Average of three determinations.
[b] Reagent grade chemical.
Reprinted by permission, J. W. Mitchell, C. L. Luke and W. R. Northover, *Anal. Chem.*, **45**, 1506 (1973). Copyright by the American Chemical Society.

$\pm 0.4\%$ recoveries of 100 ng of each metal in the presence of the isotopes ^{59}Fe, ^{60}Co, ^{64}Cu, ^{51}Cr, and ^{54}Mn, respectively. Data in Table 14 show X-ray counts for the remaining elements to be reproducible within $\pm 7\%$ or better.

The reproducibility of the X-ray method is given in Table 14. The background intensities for copper, iron, and vanadium are caused by line interference due to $L\alpha$ radiation of tungsten, presence of iron in the reagents and in the X-ray spectrograph, and interference from $K\beta$ radiation of titanium, respectively. Nevertheless, the net counts obtained for 100 ng of the seven metals show very high sensitivity and reasonably good precision. The sensitivity (cps/ng) calculated from the average of the values in Table 14 was as follows: Cu = 5, Ni = 8, Co = 5, Fe = 5, Mn = 3, Cr = 4, and V = 1. It was also established that a plot of the average counts versus the concentration of metal was linear over the range of 0–100 ng of metal.

Replicate analyses for particulate iron in 0.5 g samples of a typical ultrapure sodium carbonate showed 16, 41, 27, and 70 ng/g. The variability of the data obtained was undoubtedly due to the nonuniform distribution of particulates in the solid reagents. This variation would probably be significantly reduced by analyzing larger samples. Neutron activation analysis of a sample of sodium carbonate indicated 19 ± 8 ng/g Cu and 3.3 ± 1.8 ng/g Mn. X-Ray fluorescence data for impurities in particulates separated from a sample of the same sodium carbonate showed 17 ppb of copper and no detectable manganese, an indication that the bulk of copper contamination was derived from the particulate matter. Average results of analysis in triplicate of particulates in several commercially available pure salts are given in Table 15. The reagent grade Na_2CO_3 is easily identified by inspecting these data. It appears that iron contamination, one of the most common problems in ultrapurification, is significantly induced through fallout of airborne particulates.

It was first demonstrated by this procedure that particulate contamination of several commercially available solid reagents is largely responsible for the total trace metal content. Thus purification methods for removing soluble traces must be followed by ultrafiltration to remove particulates. Protection from the atmosphere must be maintained throughout the process.

V. SPARK SOURCE MASS SPECTROMETRY

A. INTRODUCTION

Vacuum RF spark source mass spectrometry is an extremely useful multielemental survey technique primarily because virtually all elements in the periodic table can be detected. Except for hydrogen, helium, and lithium, all elements can be recorded on a single spectrum. With this method traces of various nonmetallic ions, halogens, oxygen, nitrogen, sulfur, phosphorus, and gases can be measured, thus providing a major advantage over most multielemental methods.

Practical analyses by spark source techniques have been comprehensively discussed.[73-79] Limitations in sensitivity due to diffuse background in the emulsion, plate fogging, line background, and mass resolution, are particularly well treated by Ahearn.[79] Factors limiting accuracy—for example, transmission stability of the instrument, fulfillment of reciprocity, emulsion sensitivity dependence of the element, the pattern of ionization in the spark source, integration of ion intensity profile, and relative sensitivity of the elements—have also been discussed quite adequately.[79]

B. PREPARATION OF SAMPLES

Methods for the preparation of liquids, insulators and powders, microsamples, and miscellaneous samples for spark source mass spectrographic analysis were reviewed by Guthrie.[73] During the analysis of liquids for traces by evaporation onto known presparked surfaces, several precautions are necessary. To convey the importance of sample preparation, the procedures followed in our laboratory for the analysis of deionized, quartz-distilled water are presented in detail.[80] Two liters of freshly purified water were collected in a precleaned polyethylene bottle located in laminar-flow hood. The entire sample was transferred to a 6 liter Teflon vessel* that was preleached for 36 hr in 1:1 HNO_3 at 80°C and subsequently cleaned as described in Chapter 4. The vessel was placed on a ceramic-top plate in a clean hood, and the sample was concentrated during an 8 hr period by evaporation to a final volume of 9.5 ml. During the evaporation the water was periodically swirled to prevent collection of trace residues on the walls of the vessel as the level of the liquid receded. The pH of the concentrated sample was adjusted to 1.0 with several drops of concentrated hydrochloric acid prepared by saturation of purified water with filtered HCl gas. A blank water sample was prepared by adjusting the pH of a freshly distilled sample (9.5 ml) as described previously for the concentrated sample. A comparison standard was prepared by adding known quantities of Mn, Ni, Cu, V, Co, Cr, Pt, Zn, S, Cl, Fe, and Ca from 1 $\mu g/g$ standards (0.1 M in HCl) and diluting to appropriate volumes with freshly purified water.

A syringe with a Teflon barrel, plunger, and needle was used to sample the blank, the concentrated sample, and the standard. The syringe and needle were washed and rinsed after each sample was dispensed. A known aliquot of each solution (0.01–0.05 ml) was evaporated onto the end of a tantalum plate, which served as one of the electrodes. The surface of the electrode had been cleaned by etching in high purity $HF-HNO_3$ and rinsing copiously with pure water. Next the electrode was dried under a heat lamp in the clean hood, placed in a small snap-cap polyethylene vial, and transferred in a covered polyethylene beaker to the laminar flow station guarding the spark source. The tip of the electrode on which the solution was to be deposited was presparked, and the electrode was returned to the clean room and placed on a Teflon sheet that was positioned under an infrared lamp. The sample was then applied dropwise from the syringe tip, and care was taken to confine each droplet to the presparked area of the electrode. Each droplet was evaporated to dryness until the entire sample had been applied to the electrode, and immediately after evaporation of the last droplet, the blank, standard, and concentrated sample were sparked and the spectrum recorded on photographic plates.

Data reported in Table 9 Chapter 5 demonstrate that the blanks due to trace elements present in the tantalum electrode, the HCl used to adjust pH, or the purified water were negligible, and that all the elements in the doped sample were detected. The semiquantitative results reported for traces in the purified water are thus reliable enough to distinguish between 0.1, 10, or 100 ng/g of impurities.

The very high sensitivity based on the detection of traces in the doped

* Chemplast, Inc., Wayne, N.J.

sample was unexpected. Several analyses of solutions containing indium at concentration in the range 10^{-12}–10^{-10} g showed a linear dependence of optical density with concentration. Another procedure has been reported for analyzing liquids by evaporating a drop that is confined to the tip of an electrode in an electric field.[81] Ahearn points out that this method is applicable not only to aqueous samples but to any liquid that evaporates more rapidly than the dissolved impurities in the liquid. Thus applications for determining traces in laboratory chemicals should be possible.

During the analysis of bulk solids, contamination problems are at a minimum if the sample can be used directly as one of the electrodes. Pressing pellets or electrodes from powders or insulating solids after grinding and mixing with conducting materials is fraught with difficulties. Fortunately spark source mass spectroscopy can often be used to measure the blank directly; therefore the effect of contamination on the analysis can be ascertained.

Practical applications of spark source mass spectrography are numerous.[82-89] Since the introduction of electrical detection systems and computerized data handling, the method has become a versatile analytical tool.

VI. EMISSION SPECTROSCOPY (ES)

A. GENERAL

Emission spectroscopy, an excellent tool for rapid survey analysis, provides high sensitivity for approximately 80 of the first 95 elements.[90] Because of the primary advantages of detecting economically a large number of elements during a single exposure, the method is preferred in several industrial laboratories.[91] The applicability of the method has been broadened by the development of a variety of excitation systems. Conventional dc arc excitation, hollow cathode, gas discharge sources, excitation in the vacuum ultraviolet region, and laser microprobe excitation provide measurement of most elements in the periodic table.

B. SENSITIVITY

Limits of detection estimated for the more common elements by direct dc arc excitation are reported in Table 16. Detection limits have been compared for various excitation methods including open dc arc, artificial atmosphere dc arc, high current impulse argon arc, temperature-buffered argon arc, plasma emission, high voltage arc spark, rotating disk, and copper spark.[92] Hollow-cathode tubes can be used advantageously because of their extreme sensitivity. The advent of induction-coupled plasma excita-

TABLE 16. Limits of Detection for Various Elements by Emission Spectrography[a]

Limits			
0.1–10 μg	1–10 μg	10–100 μg	10^2–10^3 μg
Ag	Al	As	Hf
Au	B	Ce	Hg
Ba	Bi	Er	P
Be	Cd	Eu	Te
Ca	Co	Ho	Th
Cs	Cr	Mo	U
Cu	Cy	Nd	W
Fe	Ga	Pr	
K	Gd	Re	
Li	Ge	Sm	
Lu	In	Ta	
Mg	Ir	Tb	
Mn	La	Zn	
Na	Nb		
Pb	Ni		
Rb	Os		
Sc	Pd		
Sn	Pt		
Sr	Rh		
V	Ru		
Y	Sb		
Yb	Si		
	Ti		
	Tl		
	Tm		
	Zr		

[a] A 2 mg sample.
Source: Ref. 90.

tion considerably brightens the future of trace analysis by spectro-chemical means. Nanogram per milliliter levels of many elements can be measured.[93-96]

C. CONTAMINATION PROBLEMS

Sample preparation is frequently required for emission spectrochemical analysis. Often samples are ground and mixed with graphite, or standards are prepared by homogenizing various oxide powders. Although quantitative results are easily affected by the blank problem, the method can be used successfully if adequate precautions are observed.[97-99]

Since concentration methods are widely used to improve detection limits of spectrographic methods, attention must be paid to the particular precautions discussed in Chapter 6. The care necessary in spectrographic analysis of mineral acids has been described.[100] Samples (100 ml) of acetic, hydrochloric, hydrofluoric, nitric, perchloric, and sulfuric acid were concentrated by a factor of 10^3. Preconcentration of traces in water-soluble inorganic salts was accomplished by carrier precipitation with 8-quinolinol, tannic acid, and thionolide according to the Mitchell procedure,[101] After purifying each reagent, indium carrier was used to obtain a 500-fold concentration of traces in potassium chloride.[101]

Ashing of samples is also frequently required. In one procedure the resin was ashed and analyzed after concentration of traces on ion-exchange resins.[102] Most biological samples are concentrated by wet or dry ashing prior to spectrographic analysis.[103] Comparison of Tables 16 and 17 shows the improved detection limits after preconcentration.

In ultratrace analysis the emission spectroscopist must be extremely well versed in ultrapurity techniques, since many of the routine techniques of ashing, grinding, mixing, and concentration are particularly susceptible to contamination problems.

TABLE 17.   Detection Limits in dc Arc Spectrography After Preconcentration

Element	Limits (μg/g)		
	Ashing[a]	Collection[b]	Evaporation[c]
Calcium	1	0.01	0.0001
Chromium	0.1	0.002	0.001
Cobalt	0.1	0.01	0.001
Copper	0.01	0.001	0.0001
Iron	0.02	0.002	0.001
Lead	0.2	0.02	0.001
Magnesium	0.002	0.0002	0.00005
Nickel	0.1	0.02	0.001
Sodium	2	—	0.001
Zinc	2	0.2	0.001

[a] 50-fold concentration.
[b] 500-fold concentration via use of 8-quinolinol–thionalide–tannic acid system with indium.
[c] 10,000-fold concentration by evaporation of an acid.

VII. ATOMIC ABSORPTION (AAS) AND ATOMIC FLUORESCENCE SPECTROSCOPY (AFS)

Atomic absorption and fluorescence spectroscopy based on nonflame atomization cells are highly sensitive techniques for ultratrace analysis. Table 18 presents results for sensitivity obtained by L'vov.[104] The amounts of different elements actually measured are given, along with the corresponding experimental conditions used. Sensitivity results obtained by extrapolation to the absolute amount of each element that would produce 1% absorption in a 2.5 mm diameter graphite tube are also reported.

Atomic fluorescence signals measured for 35 elements in flame and nonflame devices showed very low limits of detection for some elements.[105] With spectral discharge lamps sensitivities of 2×10^{-10} g for cadmium and zinc have been reported.[106] Good calibration curves for 10^{-7} M solution of cadmium and 5×10^{-8} M Zn were obtained.[107, 108] Over the range 10^{-5}–10^{-7} M, investigators found no detectable interference from 41 cations and 18 anions at hundred fold concentrations relative to zinc and cadmium. Near-linear curves over the range 1×10^{-4} to 10 μg/ml for zinc, 2×10^{-4} to 1.0 μg/ml for cadmium, 0.1 to 10^3 μg/ml for mercury, and 4×10^{-2} to 10^3 μg/ml for thallium were reported using spectral discharge lamps.[109] In a comparison of these data with detection limits of other analytical methods, the size of sample capable of being analyzed must be considered. Since 2–10 μl samples are routinely injected in graphite tubes for AAS, a sensitivity of 10^{-12}g indicates that the original sample must contain more than 10^{-9} g/ml of the element. Although various controversies over the relative sensitivities of flame emission (nitrous oxide-acetylene flame), AAS, and AFS have occurred,[110] it is generally agreed that flameless methods are more sensitive by an order of magnitude or greater for most elements. Because flameless methods are highly specific, extremely sensitive, relatively simple, and economic, they should find widespread application, provided their simplicity and reliability can be made comparable to flame methods.

Several limitations of the methods exist. In most cases samples must be dissolved before analysis is possible, and matrix effects can be considerable. Thus standards closely matching the composition of the sample must be prepared and blank problems, as is the case for most highly sensitive methods, must be overcome. Restriction to liquid no longer exists, however, since methods have been developed recently for the direct analysis of solids.[111, 112] By fashioning an electrode from the sample and using it as the cathode of a low pressure discharge, atomic vapor is produced by cathodic sputtering. This method should be applicable to the analysis of a wide range of metals and alloys.[113]

TABLE 18. Sensitivity Data with L'vov Furnace

Element, Line (nm)	Tube Diameter (mm)	Argon Pressure (atm)	Temperature (°C)	Measured Amount (g)	Sensitivity (g/1% absorption)
Ag 328.1	2.5	2	1800	5×10^{-13}	1×10^{-11}
Al 309.3	4.5	1	2100	2.5×10^{-11}	1×10^{-12}
Au 242.8	2.5	2	1700	7×10^{-11}	1×10^{-11}
B 259.8	2.5	2	2400	5×10^{-9}	2×10^{-10}
Ba 553.5	3.0	1	2200	1×10^{-10}	6×10^{-11}
Be 234.9	4.5	6	2400	2.6×10^{-12}	3×10^{-11}
Bi 306.8	2.5	2	1800	2.5×10^{-11}	4×10^{-12}
Ca 422.7	2.5	2	2300	2.5×10^{-11}	4×10^{-10}
Cd 228.8	1.2	1	1500	6×10^{-14}	8×10^{-11}
Co 240.7	2.5	2	2200	7.5×10^{-12}	2×10^{-10}
Cr 257.9	2.5	2	2200	5×10^{-11}	2×10^{-11}
Cs 852.1	2.5	2	1900	6.6×10^{-12}	4×10^{-10}
Cu 324.7	2.5	2	2100	6.3×10^{-12}	6×10^{-12}
Fe 248.3	2.5	2	2100	2.5×10^{-11}	1×10^{-11}
Ga 287.4	2.5	2	2100	2.5×10^{-11}	1×10^{-11}
Hg 253.7	2.5	2	700	4.5×10^{-10}	8×10^{-11}
In 303.9	2.5	2	1900	8×10^{-12}	4×10^{-10}
K 404.4	2.5	2	1800	6.3×10^{-10}	4×10^{-11}
Li 670.8	3.0	1	1900	5×10^{-11}	3×10^{-11}
Mg 285.2	4.5	2	1800	3×10^{-12}	4×10^{-11}
Mn 279.5	2.5	2	2000	2.5×10^{-12}	2×10^{-11}
Mo 313.3	2.5	2	2500	5×10^{-12}	3×10^{-11}
Ni 232.0	2.5	2	2200	2.5×10^{-11}	9×10^{-11}
Pb 283.3	2.5	2	1900	3×10^{-11}	2×10^{-11}
Pd 247.6	2.5	2	2100	5×10^{-11}	4×10^{-11}
Pt 265.9	2.5	2	2300	2.5×10^{-10}	1×10^{-11}
Rb 780.0	2.5	2	1900	7.5×10^{-12}	1×10^{-11}
Rh 343.5	2.5	2	2300	6.3×10^{-11}	8×10^{-11}
Sb 231.1	2.5	2	2000	5×10^{-11}	5×10^{-11}
Se 196.1	2.5	2	1600	2×10^{-10}	9×10^{-11}
Si 251.6	2.5	2	2250	2.7×10^{-12}	5×10^{-11}
Sn 286.3	2.5	2	2000	1×10^{-11}	2×10^{-11}
Sr 460.7	3.0	1	2200	2×10^{-11}	1×10^{-11}
Te 214.3	2.5	2	2000	7.6×10^{-12}	1×10^{-11}
Ti 365.3	2.5	2	2500	5×10^{-10}	4×10^{-11}
Tl 276.8	2.5	2	1800	2.5×10^{-12}	1×10^{-11}
Zn 213.8	4.5	4	1500	1×10^{-12}	3×10^{-11}

Source: L'vov.[104]

To avoid systematic errors, there can be no high temperature, gas phase reactions of trace elements with constituents in the matrix or with other components present in the sample. A frequently unrecognized source of serious error is the possible formation of nonvolatile compounds, such as spinels, carbides, nitrides, or phosphates, by high temperature reaction of the element with the cell material.[114]

The introduction of multielement lamps and flameless atomization cells for AAS and AFS has made these techniques quite useful for practical applications in ultratrace analysis. However the well-founded speculation of T. S. West, ". . . the sensitivity and specificity of AFS will yet make it one of the most widely used and revolutionary techniques of our time,"[106] has still not been realized. Tremendous advances have been made nevertheless, and practical applications of the methods are increasing annually.

VIII. SUMMARY

No single analytical technique can be labeled as "the" most capable method for ultratrace analysis. Though each method will meet some of the criteria for an ideal method (Table 19), no technique has all these capabilities.

Consequently a combination of the broad range of methods available to the analytical chemist (Tables 1 and 2, Chapter 2) must be used to complement one another. It is therefore unlikely that one method will emerge in the near future to a level of distinct superiority over all others. Instead, the well-established tendency for each analytical chemist to perfect and develop his specialty until it becomes uniquely suited for certain types of analytical problems is likely to be continued.

TABLE 19. **Criteria for an Ideal Analytical Method**

1. Extremely sensitive for many elements.
2. Highly specific, but capable of simultaneous multielemental determinations.
3. Nondestructive for solids, liquids, and gases.
4. Independent of matrix effects and chemical interferences.
5. Free from contamination problems.
6. Inexpensive.
7. Simple to operate.
8. Capable of automatic operation.
9. Capable of giving absolute values independent of standards.
10. Highly precise and accurate.
11. Reasonably independent of analyst error.

REFERENCES

1. G. Hevesy and H. Levi, *Nature*, **136**, 103 (1935).
2. *Activation Analysis: A Bibliography Through 1971*, G. J. Lutz, R. J. Boreni, R. S. Maddak and W. W. Meinke, eds. National Bureau Standards Technical, Note 467, Government Printing Office, Washington, D.C., 1971.
3. E. N. Jenkins and A. A. Smales, *Quart. Rev.* **10**, 83 (1956).
4. T. L. Izenhour and G. H. Morrison, *Anal. Chem.*, **36**, 1089 (1964).
5. F. Girardi, G. Guzzi, and G. DiCola, *J. Radioanal. Chem.*, in press.
6. G. C. Goode, C. W. Baker, and N. M. Brooke, *Analyst*, **728**, September 1969.
7. M. Rakovic, in D. Cohen, ed., *Activation Analysis*, CRC Press, Cleveland, 1970, p. 67.
8. M. Simkova, J. Slunecko, and F. Kukula, *Chem. Zvesti*, **19**, 115 (1964).
9. K. Kudo and N. Suzuki, *J. Radioanal. Chem.*, **19**, 55 (1974).
10. J. Starý and J. Ružička, *Talanta*, **18**, 1 (1971).
11. N. K. Baishya and R. B. Heslop, *Crit. Rev. Anal. Chem.*, **2**, 345 (1971).
12. J. W. Mitchell and R. Ganges, *Anal. Chem.*, **46**, 503 (1974).
13. J. W. Mitchell and R. Ganges, *Talanta*, **21**, 735 (1974).
14. J. W. Mitchell and D. L. Shanks, paper 173, presented at ACS National Meeting, Los Angeles, April 1974.
15. J. E. Riley and J. W. Mitchell, paper 270, presented at Pittsburgh Conference, Cleveland, March 1974.
16. O. U. Anders, *Anal. Chem.*, **36**, 564 (1964).
17. O. T. Hogdahl, Symposium on Radiochemical Methods and Analysis, Salzburg, 1964.
18. M. Okada, *Anal. Chem.*, **45**, 1578 (1973).
19. G. H. Morrison and N. M. Potter, *Anal. Chem.*, **44**, 839 (1972).
20. J. W. Mitchell and W. R. Northover, Bell Telephone Laboratories Technical Memorandum, January 3, 1972.
21. J. W. Mitchell, J. E. Riley, and W. R. Northover, *J. Radioanal. Chem.*, **18**, 133 (1973).
22. J. W. Mitchell, C. L. Luke, and W. R. Northover, *Anal. Chem.*, **45**, 1503 (1973).
23. W. W. Meinke, International Congress on Analytical Chemistry, IUPAC Kyoto, April 1972.
24. R. C. Koch, *Activation Analysis Handbook*, Academic Press, New York, 1960.
25. W. S. Lyon, ed., *Guide to Activation Analysis*, Van Nostrand, Princeton, N.J., 1964.

26. C. E. Cronthamel, *Applied Gamma Ray Spectrometry*. Pergamon Press, New York, 1960.

27. W. W. Meinke, *Anal. Chem.*, **32**, 104 (1960).

28. G. W. Leddicotte, *Anal. Chem.*, **36**, 419R (1964).

29. H. J. M. Bowen, and D. Gibbons, *Radioactivation Analysis*, Clarendon Press, Oxford, 1963.

30. A. A. Smales, in *Trace Characterization*, National Bureau of Standards Monograph 100, W. W. Meinke and B. F. Scribner, eds., Government Printing Office, Washington, D.C., 1967, pp. 307-336.

31. National Bureau of Standards Technical Note 501, pp. 106-114, 1969.

32. National Bureau of Standards Technical Note 404, pp. 143-186, 1966.

33. J. Ružička and J. Starý, *Talanta*, **11**, 691-696 (1964).

34. J. Starý, J. Ružička, and M. Salamon, *Talanta*, **10**, 375-381 (1963).

35. D. A. Beardsley, G. Briscoe, J. Ružička, and M. Williams, *Talanta*, **14**, 879-886 (1967).

36. J. Ružička, *Coll. Czech. Chem. Commun.*, **10**, 1808-1812 (1965).

37. J. Starý and J. Ružička, *Talanta*, **11**, 697-792 (1964).

38. J. Ružička and J. Starý, *Substoichiometry in Radiochemical Analysis*, Pergamon Press, New York, 1968.

39. J. Starý, J. Ružička, and M. Salamon, *Talanta*, **10**, 375 (1963).

40. R. Hill, A. G. Jones, and D. E. Palin, *Chem. Ind.* (*London*), **162**, February 6, 1954.

41. P. Sorensen, *Anal. Chem.*, **28**, 1318 (1956).

42. J. K. Whitehead, *Biochem. J.*, **68**, 662 (1958).

43. W. R. Shields, ed., *Technical Note* 546, 1970, p. 34.

44. H. Müller, *Z. Anal. Chem.*, **113**, 161 (1938).

45. R. Alverez, P. J. Paulsen, and D. E. Kelleher, *Anal. Chem.*, **41**, 944 (1969).

46. F. D. Leipzigen, *Anal. Chem.*, **37**, 171 (1965).

47. B. R. Kowalski, T. L. Isenhour, and R. E. Sievers, *Anal. Chim. Acta*, to be published.

48. A. E. Jenkins and J. R. Majer, *Talanta*, **14**, 777 (1967).

49. A. E. Jenkins, J. R. Majer, and M. J. A. Reade, *Talanta*, **14**, 1213 (1967).

50. E. C. Kuehner, R. Alvarez, P. J. Paulsen, and T. J. Murphy, *Anal. Chem.*, **44**, 2050 (1972).

51. I. L. Barnes et al., *Anal. Chem.*, **45**, 880 (1973).

52. I. L. Barnes, T. J. Murphy, J. W. Gramlich, and W. R. Shields, *Anal. Chem.*, **45**, 1881 (1973).

53. P. J. Paulsen, R. Alvarez, and C. W. Mueller, *Anal. Chem.*, **42**, 673 (1970).

54. W. T. Grubb and P. D. Zemany, *Nature*, **176**, 221 (1955).

55. C. L. Luke, *Anal. Chem.,* **35,** 1551 (1963).

56. C. L. Luke, *Anal. Chem.,* **36,** 318 (1964).

57. W. T. Campbell, E. F. Spono, and T. E. Green, *Anal. Chem.,* **38,** 987 (1966).

58. A. Volborth, Nevada Bureau of Mines, Report 6, Part A, 1963.

59. R. A. Jones, *Anal. Chem.,* **31,** 1341 (1959).

60. P. K. Koh and B. J. Caugherly, *J. Appl. Phys.,* **23,**427 (1952).

61. H. J. Beattie and R. M. Brissey, *Anal. Chem.,* **26,** 980 (1954).

62. W. J. Campbell and J. D. Brown, *Anal. Chem.,* **36,** 312R (1964).

63. B. J. Mitchell, *Anal. Chem.,* **30,** 1894 (1958).

64. T. C. Loomis, *Ann. N.Y. Acad. Sci.,* **137,** 284 (1966).

65. E. F. Kaelable, *Handbook of X-Rays,* McGraw-Hill, New York, 1967, Chapter 37.

66. C. L. Luke, *Anal. Chim. Acta.,* **41,** 237–250 (1968)

67. T. C. Loomis and K. H. Storks, *Bell Laboratories Record,* **45,** 2–7 (1967).

68. J. E. Kessler and S. M. Vincent, paper 70 presented at the Pittsburgh Conference, Cleveland, March 1972.

69. E. Maienthal and R. A. Paulson, National Bureau of Standards Technical Note 545, 1970, p. 53.

70. D. H. Freeman and W. L. Zielinski, Jr., National Bureau of Standards Technical Note 549, 1971, p. 62.

71. J. W. Mitchell, C. L. Luke, and W. R. Northover, *Anal. Chem.,* **45,** 1503 (1973).

72. C. L. Luke, T. Y. Kometani, J. E. Kessler, and T. C. Loomis, *Environ. Sci. Technol.,* **6,** 1105 (1972).

73. J. W. Guthrie, in *Mass Spectroscopy in Physics Research,* National Bureau of Standards Circular 522, Government Printing Office, Washington, D.C., 1953, Chapter IV, p. 111.

74. E. B. Owens and N. A. Giardino, *Anal. Chem.,* **35,** 1172 (1963).

75. N. B. Hannay and A. J. Ahearn, *Anal. Chem.,* **26,** 1056 (1954).

76. N. B. Hannay, *Science,* **134,** 1220 (1961).

77. A. J. Ahearn, ed., *Mass Spectrometric Analysis of Solids,* Elsevier, New York, 1966.

78. A. J. Ahearn, Bell Laboratories Technical Memorandum, Murray Hill, N.J., May 7, 1963.

79. A. J. Ahearn, in *Trace Characterization, Chemical and Physical,* W. W. Meinke and B. F. Scribner, eds., National Bureau Standards Monograph 100, Government Printing Office, Washington, D.C., April 28, 1967.

80. D. L. Malm and J. W. Mitchell, unpublished work.

81. A. J. Ahearn, *J. Appl. Phys.,* **32,** 1197 (1961).

82. M. S. Chupakhin, J. A. Kagakov, and O. I. Kryuchkova, *J. Anal. Chem. USSR*, **24**, 1-S (1969).

83. E. B. Owens, *Adv. Mass Spectrosc.*, **3**, 101 (1966).

84. J. Roboz, *Introduction to Mass Spectrometry*, Interscience, New York, 1968.

85. F. Aulinger, *Z. Anal. Chem.*, **221**, 70 (1966).

86. G. H. Morrison, J. T. Gerard, A. T. Kashuba, E. U. Gangadharam, A. M. Rothenberg, N. M. Potter, and G. B. Miller, *Geochim. Cosmochim. Acta, Suppl. I.*, **4**, 383 (1970).

87. G. H. Morrison and A. T. Kashuba, *Anal. Chem.*, **41**, 1842 (1969).

88. J. Franzen and K. D. Schuy, *Z. Anal. Chem.*, **225**, 295 (1967).

89. A. D. Wilson, *Analyst (London)* **89**, 18 (1964).

90. D. L. Nash, Bell Telephone Laboratories Memorandum, April 8, 1970.

91. A. J. Barnard, E. F. Joy, K. Little, and J. D. Brooks, *Talanta*, **17**, 785 (1970).

92. E. C. Snooks, in *Ultrapurity*, M. Zief and R. Speights, eds., Dekker, New York, 1972, Chapter 18, p. 437.

93. American Society for Testing and Materials, *Methods for Emission Spectrochemical Analysis*, 4th ed., ASTM, Philadelphia, 1964, p. 108.

94. ASTM Special Technical Publication 76, Philadelphia, 1948.

95. C. D. West and D. N. Hume, *Anal. Chem.*, **36**, 412 (1964).

96. G. W. Dickinson and V. A. Fassel, *Anal. Chem.*, **41**, 1021 (1969).

97. J. D. Nohe, *Appl. Spectrosc.*, **21**, 364 (1967).

98. I. P. Alimarin, ed., *Analysis of High-Purity Materials*, 1968 Israel Program For Scientific Translations, Jerusalem, 1968.

99. A. J. Barnard, E. F. Joy, K. Little, and J. D. Brooks, *Talanta*, **17**, 785 (1970).

100. N. A. Kerschner, E. F. Joy, and A. J. Barnard, *Appl. Spectrosc.* **25**, 542 (1971).

101. C. Farquhar et al., *Anal. Chem.*, **38**, 208 (1966).

102. F. T. Birks et al., *Analyst (London)* **89**, 36 (1964).

103. G. H. Morrison et al., *Appl. Spectrosc.*, **23**, 349 (1969).

104. B. V. L'vov, *Spectrochim. Acta*, **24B**, 53 (1969).

105. J. A. Dean and T. C. Rains, *Flame Emission and Atomic Absorption Spectrometry*, Vol. 1, *Theory:* Vol. 2, *Components and Techniques*, New York, 1969, 1971.

106. T. S. West, in *Trace Characterization*, W. W. Meinke and B. F. Scribner, eds., National Bureau of Standards Monograph 100, Government Printing Office, Washington, D.C., 1967, pp. 215–306.

107. R. M. Dagnall, T. S. West, and P. Young, *Talanta*, **13**, 803 (1966).

108. R. M. Dagnall, K. C. Thompson, and T. S. West, *Anal. Chim. Acta*, **36**, 269 (1966).

109.   C. Veillon, J. M. Mansefield, M. L. Parsons, and J. D. Winefordner, *Anal. Chem.*, **38**, 205 (1966).

110.   G. P. Christian, *Appl. Spectrosc.*, **25**, 660 (1971).

111.   B. M. Gatehouse and A. Walsh, *Spectrochim. Acta*, **16**, 602 (1960).

112.   D. S. Gough, P. Hannaford, and A. Walsh, *Spectrochim. Acta*, **B28**, 197 (1973).

113.   A. Walsh, *Anal. Chem.*, **46**, 698A (1974).

114.   G. Tölg, *Talanta*, **21**, 327 (1974).

APPENDIX

I. ABBREVIATIONS, DEFINITIONS, AND SYMBOLS

AAS, AFS	atomic absorption and fluorescence spectroscopy
Class 100	maximum number of particles 0.5 μm and larger per cubic foot of air = 100
Coprex	X-ray analysis of elements separated and preconcentrated on paper disks via coprecipitation with an organic precipitant
D.L.	detection limit
ES	emission spectroscopy
FW	formula weight
H(fod)	1,1,1,2,2,3,3-heptafluoro-7,7-dimethyl-4,6-octanedione
HAP	hydrated antimony pentoxide
HD	dithizone
HEPA	high efficiency particulate air
HPDC	1-pyrrolidinedithiocarbamic acid*
HTTA	thenoyltrifluoroacetone
L.F.	laminar flow
NAA	neutron activation analysis
RID	radioisotope dilution
SID	stable isotope dilution
SRID	substoichiometric radioisotope dilution
TOPO	tri-n-octylphosphine oxide
trace	1–100 μg/g
ultratrace	< 1 μg/g
vitreous silica	glassy modification of silica; transparent vitreous silica ("fused quartz") is prepared by fusing quartz crystal
$\bar{x}$	average or mean
σ_{bl}	standardard deviation of the blank

* HPDC occurs in the literature even though the correct terminology should be HPCD, for 1-pyrrolidinecarbodithioic acid.

II. STORAGE CONDITIONS FOR MINIMUM LOSS OF TRACE ELEMENTS

Element	Concentration	Container	pH or Medium	Period of Stability	Reference (Chapter 2)
Ag	—	Polyethylene	≤1.5, seawater	—	45
	1.0 μg/ml	Borosilicate	0.1 M NH$_4$OH	30 days	56
	0.05 μg/ml	Borosilicate	0.1 M NH$_4$OH	30 days	59
		Polyethylene, Vycor, Teflon	NaCl	30 days	59
	2 × 10^{-6} M	Teflon	—	1 day	57
Al	1.0 μg/ml	Borosilicate	1.5	4 weeks	46
Au	1.0 μg/ml	Borosilicate	2.0	4 weeks	47
Ba-La	—	Borosilicate	—	—	43
Bi	1.0 μg/ml	Borosilicate	2.0	4 weeks	47
Ca	0.5 μg/ml	Borosilicate	1.5	4 weeks	46
Cd	0.2 μg/ml		≤2.0	4 weeks	47
Ce	—	Borosilicate	—	—	43
Co	—	Polyethylene	Seawater	60 days	45
	1.0 μg/ml	Borosilicate	1.5	4 weeks	46
Cr	0.05 ng/ml	Borosilicate, polyethylene	—	30 weeks	58
	1.0 μg/ml	Borosilicate	1.5	4 weeks	46
Cs	—	Borosilicate	—	—	43
	—	Polyethylene	Seawater	30 days	45
Cu	1.0 μg/ml	Borosilicate	1.5	4 weeks	46
Fe	—	Polyethylene	Seawater	55 days	45
	1.0 μg/ml	Borosilicate	1.5	4 weeks	46
Hf	—	Polyethylene	≤1.5, seawater	—	45
Hg	10^{-8} – 10^{-5}M	Polyethylene	O, natural water in		48, 49, 55
	0.1 μg/ml	All containers	HCl, HNO$_3$ H$_2$SO$_4$	1 week	52, 53
I	—	Borosilicate	—	—	43
In	1.0 μg/ml	Borosilicate	2.0	4 weeks	47
	—	Polyethylene	≤1.5, seawater	60 days	45
Li	0.2 μg/ml	Borosilicate	2.0	4 weeks	59
Mg	0.5 μg/ml	Borosilicate	1.5	4 weeks	46
Mo	1.0 μg/ml	Borosilicate	1.5	4 weeks	46
Ni	1.0 μg/ml	Borosilicate	1.5	4 weeks	46
Pb	1.0 μg/ml	Borosilicate	1.5	4 weeks	46
Pd	1.0 μg/ml	Borosilicate	2.0	4 weeks	47
Pt	1.0 μg/ml	Borosilicate	2.0	4 weeks	47
Rh	1.0 μg/ml	Borosilicate	2.0	4 weeks	43, 47
Ru	1.0 μg/ml	Borosilicate	2.0	4 weeks	47
Se	1.0 μg/ml	Borosilicate	2.5–7 (HNO$_3$)	15 days	61
Sb	1.0 μg/ml	Polyethylene, borosilicate	< 1.5, seawater 2.0	55 days 4 weeks	45, 47
Sr	1.0 μg/ml	Borosilicate	1.5	4 weeks	46
Ti	1.0 μg/ml	Borosilicate	1.5	4 weeks	46
Te	1.0 μg/ml	Borosilicate	2.0	4 weeks	47
U	—	Polyethylene	Seawater	20 days	45
V	1.0 μg/ml	Borosilicate	1.5	4 weeks	46
Zn	0.5 ng/ml	Borosilicate	1.5	30–60 days	59

III. SUPPLIERS OF SPECIALTY PRODUCTS FOR THE ANALYTICAL LABORATORY

A. LAMINAR-FLOW HOODS

Air Control, Inc., Norristown, Pa. 19401
Agnew-Higgins, Inc., Garden Grove, Calif. 92641
Baker Company, Inc., Biddeford, Me. 04005
Contamination Control, Inc., Kulpsville, Pa. 19443
Edcraft Industries, Inc., Linden, N.J. 07036
Envirco, P.O. Box 6468, Albuquerque, N.M. 87107
Environmental Air Control Inc., Hagerstown, Md.
Environmental General Corp., Alexandria, Va. 22304
Laminaire Corp., Rahway, N.J. 07065

B. VITREOUS SILICA

Amersil, Inc., Hillside, N.J.
Corning Glass Works, Corning, N.Y. 14830
Dynasil Corp. of America, Berlin, N.J. 08009
General Electric Co., Willoughby, Ohio 44094
Heraeus-Schott Quarzschmelze, GmbH, West Germany.
Quartz Products Corp. (subsidiary of Quartz & Silice, Paris), Plainfield, N.J.
Thermal American Fused Quartz Co. (Subsidiary of Thermal Syndicate Ltd., England), Montville, N.J.

C. GLASS

Ace Glass Inc., Vineland, N.J.
Corning Glass Works, Corning, N.Y.
Kimble Glass Division, Owens-Illinois Inc., Vineland, N.J.
Kontes Glass Co., Vineland, N.J.
SGA Scientific Inc., Bloomfield, N.J.
Wheaton Glass Co., Inc., Millville, N.J.

D. PLASTICS

1. Fluorocarbons

Chemplast Inc., Wayne, N.J.

2. Polyethylene, Polypropylene, and Selected Fluorocarbon Ware

Bel-Art Products, Pequannock, N.J.
Nalge Co., Rochester, N.Y.

E. HIGH PURITY CHEMICALS AND METALS

Alfa Inorganics, Beverly, Mass.
Apache Chemicals, Rockford, Ill.
Asarco, S. Plainfield, N.J.
Atomergic Chemicals, Long Island, N.Y.
Brinkman Instruments Inc., Westbury, N.Y.
Cominco American Inc., Spokane, Wash.
Englehart Industries, Newark, N.J.
Fischer Scientific Co., Pittsburgh, Pa.
Imanco, Monsey, N.Y.
J. T. Baker Chemical Co., Phillipsburg, N.J.
Johnson Matthey, England
Kerr McGee Corp., Oklahoma City, Okla.
Mallinckrodt Chemical Works, St. Louis, Mo.
Materials Research Corp., Orangeburg, N.Y.
Merck, Darmstadt, Germany
National Bureau of Standards, Gaithersburg, Md.
Spex Industries, Metuchen, N.J.
Texas Instruments, Inc., Dallas, Tex.

F. ION–EXCHANGE RESINS[a]

Activity Group	DUOLITE[b]	DOWEX[c]	IONAC[d]	AMBERLITE[e]
Cation exchangers				
Gel type				
$-SO_3Na$	C-20	50	C-240	IR-120
$-SO_3Na$	C-20X10	50-X10	C-250	IR-122
$-SO_3Na$	C-20X12	50-X12	C-255	IR-124
$-SO_3Na$	C-25D			
$-COOH$	C-433	CCR-2	C-270	IRC-84
$-P(O)(OH)_2$	ES-63			
Macroporous				
$-SO_3Na$	ES-26	MSC-1		200
$-SO_3Na$				252
$-COOH$	CC-4			IRC-50
Anion exchangers				
Gel type				
$-NH(R)$	A-2	3	A-315	IR-45
$-N(R)_3 + (1)$	A-101D, ES-109	1, 21K	A-540	IRA-400, -402
$-N(R)_3 + (2)$	A-102D	2	A-550	IRA-410
Macroporous				
$-NH(R)$	ES-368			IRA-94
$-N(R_3) + (1)$	A-161			IRA-900
$-N(R_3) + (2)$				IRA-910
Chelating resins				
$-N \Big\langle \begin{array}{l} CH_2COOH \\ CH_2COOH \end{array}$				XE-318
$-C \Big\langle \begin{array}{l} NH_2 \\ N\text{-}OH \end{array}$	CS-346			
Mixed-bed resins				
$-SO_3H$ $+$ $-N(R)_3 + (1)$				MB-1

[a] Every manufacturer produces a multitude of variations with respect to particle size, cross-linking, and chemical structure. This table lists the most important resins.
[b] Diamond Shamrock Chemical Co., Redwood City, Calif.
[c] Dow Chemical Co., Midland, Mich.
[d] Ionac Chemical Co., Birmingham, N.J.
[e] Rohm and Haas Co., Philadelphia.

G. FILTERS

Filter type	Manufacturer Millipore[a]	Gelman[b]	Nuclepore[c]	Selas[d]	Application
Membrane					
Cellulose triacetate	Celotate (0.2–10)	Triacetate Metrice (0.2–5.0)			Low molecular weight alcohols
Cellulose (mixed esters)	MF-Millipore (0.025–8.0)				For general use except ketones, esters, strong acids, and bases
Acrylonitrile-polyvinyl chloride copolymer		Acropor (0.2–5.0)			For general utility except ketones, acids, methanol, and methylene chloride
Polytetrafluoro-ethylene	Fluoropor				For general use except strong bases and halogenated hydrocarbons
Polycarbonate			Nuclepore (0.03–8.0)		
Metal					
Silver				Flotronics (0.2–5)	

[a] Millipore Corp., Bedford, Mass.
[b] Gelman Instrument Co., Ann Arbor, Mich.
[c] Nuclepore Corp., Pleasanton, Calif.
[d] Selas Flotronics, Inc., Springhouse, Pa. Pore size (μm); 0.2 μm porosity is recommended for most final filtrations.

H. LYOPHILIZERS

Company	Location
Virtis Co., Inc.	Gardiner, N.Y.
Glaseal Div., Thermovac Industries Corp.	Copague, N.Y.
Refrigeration for Science, Inc.	Island Park, N.Y.
Edwards High Vacuum Inc.	Grand Island, N.Y.
Bench Scale Equipment Co.	Dayton, Ohio
New Brunswick Scientific Co., Inc.	New Brunswick, N.J.
Labconco Corp.	Kansas City, Mo.
Ace Glass Inc.	Vineland, N.J.
FTS Systems, Inc.	Stone Ridge, N.Y.
Hull Corp.	Hatboro, Pa.
Leybold-Heraeus GmbH	West Germany

INDEX

253